[Notified in A.C.Is. 20 January, 1943]

226 / Publications / 52

NOT TO BE PUBLISHED

The information given in this document is not to be communicated, either directly or indirectly, to the Press or to any person not holding an official position in His Majesty's Service.

HANDBOOK OF ENEMY AMMUNITION

PAMPHLET No. 5

GERMAN SMALL ARMS AMMUNITION, GRENADES AND DEMOLITION CHARGES

By Command of the Army Council,

THE WAR OFFICE,
 20 *January*, 1943.

HANDBOOK
OF
ENEMY AMMUNITION
Pamphlet No. 5

CONTENTS TABLE
German Ammunition

	PAGE
Cartridges, S.A.A., 7·92-mm. (Scharfe Patronen, 7·92-mm.)... ...	3
Cartridges, pistol, 9-mm. (Pistolenpatronen)	9
Interchangeability	9
Markings on ammunition for 13-mm. to 30-mm. weapons	17
Cartridges, 2-cm., Oerlikon, Mauser and Solothurn	18
Fuze A.Z. 1501	27
Fuze A.Z. 1504	28
Fuze Z.Z. 1505 (Self-destroying) with and without delay	29
Fuze A.Z. ~~5405~~ 50.45	30
Grenade, hand, H.E., egg-shaped. (Eihandgranate)	50
Grenade, hand, H.E., stick type (Stielhandgranate)	50
Demolition charges with hollow charge	51

HANDBOOK
OF
ENEMY AMMUNITION

GERMAN CARTRIDGES, S.A.A., 7·92-mm.

The following table gives the identification markings of these Mauser-type cartridges :—

1. TYPES

German Nomenclature	Type	Identification Marking		Service
		Cartridge Base	Bullet	
Patronen :—				
s.S. ...	Ball	Green annulus	Plain	Land and Air.
l.S. ...	Practice ball	Green stripe	Plain	A.A.
S.m.E. ...	Semi A.P.	Blue annulus	—	Land.
S.m.K. ...	A.P.	Red annulus	Plain	Land and Air.
S.m.KL'spur	A.P./T.	Red annulus	Black tip	Do.
P.m.K. ...	A.P./I.	Black or red annulus or red stripe.	Plain	Air.
—	H.V.A.P.	Red annulus	Green tip	Air.
—	H.V.A.P./T.	Red annulus	Green ring below black tip.	Air.
—	H.V.A.P./I	Black annulus	Green tip	Air.
S.m.K.(H)	H.V.A.P. with tungsten carbide core.	Red annulus or cap.	Black envelope.	Land and Air.
B ...	Explosive Incendiary.	Black annulus	Chromium plated tip or black envelope with plain tip.	Air.
l.S.L'.spur	Practice Tracer.	Green stripe	Black tip	A.A.

2. MARKINGS

A comparison of these markings reveals the following facts :—

A.P. types of cartridges have a red base marking. Ball and practice types of cartridges have a green base marking. Cartridges containing an incendiary charging have a black base marking. In the case of the armour-piercing incendiary cartridge, the red to indicate A.P., or black to indicate incendiary is used.

The Semi A.P. cartridge has a blue base marking.

Cartridges containing tracing composition have a black tip to the bullet.

Cartridges containing high-velocity propellant charges have a green tip to the bullet. Where a tracer is included and in consequence the tip is black, a green ring is marked below the black tip.* The high-velocity armour-piercing cartridge with tungsten carbide core is an exception to this system of green marking.

3. CASES AND CAPS

A description of the types of cartridge cases and initiator caps used with this ammunition is given in Pamphlet No. 2. The general appearance and base stampings of a typical cartridge are shown in Fig. 1, which also gives the significance of the stampings where known. Wads between the propellant and bullet are not used.

The brass caps are secured by three stabs and contain an approximately ·44-grain filling consisting of mercury fulminate, potassium chlorate, antimony sulphide and glass. The composition is varnished on the surface and covered with a disc of metal foil, the edge of the disc being upturned round the wall of the cap. The anvil formed in the cap chamber of the case is indented into the foil disc.

4. PROPELLANT CHARGES

The propellants vary in nature and form with the various types of cartridges, but in all types it is in granular form either as flakes or tubes.

The flake propellant is graphited and is a nitrocellulose powder consisting basically of 94·97 per cent. nitrocellulose (Nitrogen content 13·22 per cent.), with diphenylamine as a stabilizer and ethyl carbamite. Potassium sulphate is also included, presumably to reduce the flash. The dimensions of the flake are ·06 × ·05 × ·04 inch. This propellant is normally used in types other than the high-velocity cartridges, the charge weight varying between 44 and 45 grains.

Two natures of graphited tubular propellant are used. One is used in the high-velocity cartridges in which the bullets bear green marking, the other is used in the high-velocity A.P. cartridge with tungsten carbide core.

The propellant used in the cartridges bearing the green marking consist basically of approximately 60 per cent. of penta-erythritol-tetra-nitrate with 32 per cent. of nitrocellulose (Nitrogen content 12·5 per cent.). The stabilizer is diphenylamine and ethyl centralite is used as a moderator. Potassium sulphate is also included, presumably to reduce the flash. The average dimensions of the tubular grains are :—length ·070 in., external diameter ·049 in., internal diameter ·011 in. The charge weight varies between approximately 52 and 53 grains and the velocity is increased by approximately 300 f.s.

The propellant used in the high-velocity A.P. cartridge with tungsten carbide core consists basically of approximately 34 per cent. of penta-erythritol-tetra-nitrate with 56 per cent. of nitrocellulose (Nitrogen content

---

* The green ring is used sometimes with a plain tipped bullet. H.V. Explosive bullets have been found marked in this way.

13·2 per cent.). The stabilizer is diphenylamine. Ethyl centralite and dinitrotoluene are included presumably to moderate the rate of burning. The average dimensions of the tubular grains are :—length ·063 in., external diameter ·047 in., internal diameter ·008 in. The charge weight is approximately 55 grains and the velocity is increased by approximately 200 f.s.

5. TABLE OF WEIGHTS IN GRAINS

Type	Cartridge	Propellant Charge		Bullet	Core	Fig. No of Bullet
		Nature	Weight			
Ball	408	N.C. flake	45	198	154	2
Prac. Ball	300	N.C. flake	44	86	42	—
Semi A.P.	393	N.C. flake	45	178	89	—
A.P.	393	N.C. flake	45	178	89	3
A.P./T.	372	N.C. flake	45	157	39	4
A.P./I.	359	N.C. flake	45	156	38	5
H.V.A.P.	393	P.E.T.N./N.C. granular tube.	53	178	88	—
H.V.A.P./T.	367	Do.	53	154	40	6
H.V.A.P./I.	372	Do.	53	157	37	—
H.V.A.P. with tungsten carbide core.	419	Do.	55	194	126	7
Explosive Incendiary.	375	N.C. flake	45	167	—	8
Prac. Tracer.	309	N.C. flake	43	93	25	9

Total weights given above are for steel cartridge cases. Weights for brass cartridge cases are 7–9 grains lighter.

6. BULLETS

With the exception of the high-velocity A.P. cartridge with the tungsten carbide core all of the types included in the table of types have streamlined bullets.

In all of the types the bullet envelope is of steel, and is coated on both the inside and the outside with gilding metal. The envelope of the A.P./I. bullet is not cannelured for the attachment of the case.

No lubricant is used on the bullets.

7. s.S. BALL (FIG. 2)

This cartridge is described in Pamphlet No. 2. Details of the weights are given in the table in this pamphlet in para. 5.

8. l.S. PRACTICE BALL

The bullet consists of the envelope containing an aluminium core.

9. S.m.E. SEMI-ARMOUR PIERCING

The bullet consists of the envelope containing a core of iron or mild steel enclosed in a lead sleeve. The core is not hardened.

10. S.m.K. ARMOUR PIERCING (FIG. 3)

This cartridge is described in Pamphlet No. 2.

11. S.m.K. L'SPUR. ARMOUR PIERCING/TRACER (Fig. 4)

The cartridge is described in Pamphlet No. 2.

Tracer fillings of the following compositions have been used :—

The priming composition is common to all colours of trace and consists of magnesium powder 50·25, barium nitrate 33·68, sodium picrate 11·56 and bakelite resin, 4·52 parts.

The red trace composition, used alone or with a green trace composition, consists of the following parts : magnesium powder 32·12, strontium nitrate 49·15 parts and bakelite resin 18·73.

The green trace composition, used alone, consists of magnesium powder 33·23, barium nitrate 48·32, black mercuric sulphide 12·58 and bakelite resin 5·87 parts.

The green trace composition used in conjunction with a red composition consists of magnesium powder 27·4, barium nitrate 50·3 and bakelite resin 22·2 parts.

12. P.m.K. ARMOUR PIERCING/INCENDIARY (Fig. 5)

The envelope contains a lead tip, a hardened steel A.P. core which is tapered towards the rear, ·5 grains of yellow phosphorus and a lead base plug. The base of the A.P. core is seated in the base plug. The envelope has a small hole ·013 in. in diameter just above the streamlined portion. The hole which is closed by fusible metal and is covered by the neck of the case provides the outlet for the phosphorus after firing. The base of the bullet is soldered.

13. HIGH VELOCITY ARMOUR PIERCING

The bullet is similar in design to the S.m.K. A.P.

14. HIGH VELOCITY ARMOUR PIERCING/TRACER (Fig. 6)

The envelope contains a hardened steel A.P. core and a tracer tube enclosed in a lead sleeve and is closed at the base by a magnesium washer under which the envelope is turned. The tracer tube is of steel and is coated on both sides with gilding metal. The tube contains a night tracing composition inserted in three concave pressings, a priming composition with coned finish and finally a dark ignition compositon with a serrated finish. The ignition composition is covered by a pink plastic disc of polymerized oil.

The trace is white and bright, becoming visible at approximately 150 yards and tracing up to 1,100 yards.

Analysis of Composition

Dark Ignition		Priming		Tracing	
	Per cent.		Per cent.		Per cent.
Magnesium ...	2·7	Magnesium ...	30·0	Magnesium ...	30·9
Barium Peroxide	0·8	Barium Peroxide	39·6	Barium Nitrate	29·5
Potassium Nitrate	50·0	Strontium Picrate	14·2	Strontium Nitrate	10·3
Sulphur ...	6·1	Sodium Oxalate	6·8	Sodium Oxalate	12·6
Charcoal	15·9	Phenolic Resin ...	9·4	Phenolic Resin	12·4
AntimonySulphide	13·9			Volatile Matter and oxidation products.	4·3
Phenolic Resin ...	10·6				

15. HIGH VELOCITY ARMOUR PIERCING/INCENDIARY
The bullet is of the same type as that described in para. 12.

16. HIGH VELOCITY ARMOUR PIERCING WITH TUNGSTEN-CARBIDE CORE (FIG. 7)
The bullet, in addition to having a tungsten-carbide core instead of one of hardened steel, differs from the S.m.K. type in having a plain base instead of being streamlined and is ·34 in. shorter in length. The A.P. core is flat based and has a blunt tip. The base of the core has a ground finish while the surface finish varies along its length. The rear part of the parallel portion is rough, the forward part is ground smooth, while the ogival head is ground and polished.

17. B. EXPLOSIVE INCENDIARY (FIG. 8)
The chromium-tipped bullet contains a charging of white phosphorus in the ogival head and a striker with detonator below the ogive. The phosphorus charging in the head is sealed by a lead plug which is recessed from the base to receive the holder containing the striker and detonator. The holder consists of a steel tube coated with gilding metal. The tube is closed at one end and has an internal step to support the detonator. The forward end of the holder is closed over the detonator by a lead plug. The detonator consists of an aluminium shell with a hole in the base. The hole is closed by an aluminium disc and the shell filled with a mixture of lead styphnate, barium nitrate, antimony sulphide, calcium silicide and tetrazene. The front end of the detonator is closed by a tightly fitting lid of aluminium. The steel striker is positioned behind the detonator and is kept in a safe position by a tightly fitting steel split collar which protrudes forward beyond the striker point. The base of the bullet is closed by a lead plug under the base of which the envelope is turned.

On impact the striker pierces the detonator which breaks up the envelope and projects the phosphorus. The phosphorus ignites spontaneously in the air.

The black bullet with plain tip is mechanically the same but differs in the incendiary charging. The charging in this type is a transparent product with the consistency and appearance of paraffin. This is ignited by friction and not solely by contact with air.

18. 1.S.L'. SPUR TRACER, PRACTICE (FIG. 9)
The envelope contains an aluminium core and a tracer tube. The tube is of steel, coated with gilding metal on both sides and contains a tracing composition which produces a yellow trace up to approximately 900 yards. The tracer composition is protected by a disc of pink plastic, the whole being kept in place by an aluminium washer over which the base of the envelope is turned.

19. PERFORATION OF ARMOUR BY A.P. TYPES
It has been established by trials that the A.P. types will defeat homogeneous hard armour plate of the following thicknesses at normal impact at 100 yards :—

A.P.	12·43 mm.
H.V.A.P.	14·15 ,,
H.V.A.P. with tungsten-carbide core	19·0 ,,
A.P./T.	8·8 ,,
H.V.A.P./T.	10·5 ,,
A.P./I.	7·52 ,,
H.V.A.P./I.	8·8 ,,

The S.A.P. bullet may be expected to defeat a similar plate of 7 to 10 mm. at 100 yards, normal impact.

20. ARMOUR PIERCING CORES

The composition and hardness of the hardened steel armour-piercing cores are varied. The diamond pyramid hardness figure is approximately 900 and typical compositions are as follows :—

Carbon	1·37	per cent.
Silicon	·1 to ·3	,,
Manganese	·2 to ·4	,,
Chromium	·2 to ·5	,,
Tungsten	1·0 to 1·2	,,
Vanadium	0 to 0·2	,,

The tungsten-carbide core has an average hardness figure of 1850. The composition of the material has been found by analysis to be as follows —

Carbon	5.4	per cent.
Silicon ... ... under	·07	,,
Chromium ... ... ,,	·1	,,
Tungsten	90·4	,,
Nickel	1·8	,,
Iron	1·0	,,
Titanium	Trace	

21. VELOCITIES

The following velocities have been obtained in this country :—

	Ball	A.P.	A.P. with tungsten, carbide core	A.P./T.	A.P.I.	Explosive Incendiary	Practice ball	H.V.A.P.	H.V.A.P./T.
German name	s.S.	S.m.K.	S.m.K.(H)	S.m.K. L'spur	P.m.K.	B=Patr.	1.S.	—	—
O.V. at 90 ft.	2510 f.s.	2620 f.s.	2860 f.s.	2720 f.s.	2740 f.s.	2670 f.s.	3050 f.s.	2814 f.s.	2936 f.s.

22. PACKAGE LABELS

Typical labels for certain natures of cartridges are shown in Figs. 10, 11 and 12. The instructions given in the last line of printing on the labels for the tracer types are to the effect that the ammunition must be kept dry and protected from jolts and falls. On the label for the B. Patrone the corresponding instructions are to the effect that the ammunition must be protected from jolts and falls.

GERMAN CARTRIDGES, PISTOL, 9-mm.

The following types are known to exist :—
Cartridge, Ball, 08 (Pistolpatrone 08).
Cartridge, Semi-Armour Piercing.

BALL CARTRIDGE (FIG. 13)

This type is identified by the plain bullet.

The cartridge case is of drawn brass and is of the rimless type with an anvil and two flash channels in the cap chamber. The base of the case is stamped with a letter indicating the manufacturer, S* indicating a brass case, a number indicating the delivery number (presumably the lot number) and last two figures of the year of manufacture of the case.

The cap differs from the No. 88, used in the 7·92-mm. cartridge in dimensions and in having a smaller filling. The cap annulus is lacquered black.

The propellant charge consists of 5·6 grains of nitrocellulose powder in the form of greenish cylindrical flakes of about ·002 to ·003 inch in diameter.

The bullet consists of a lead core enclosed in an envelope of steel which is coated with gilding metal. The weight of the bullet is 123 grains and that of the core is approximately 78 grains. The bullet will pierce a steel helmet at 20 yards, normal impact.

The weight of the cartridge is approximately 190 grains and the length ·62 inches. According to a German table the velocity is 1,050 f.s.

A typical package label for this type is shown in Fig. 14.

SEMI-ARMOUR PIERCING CARTRIDGE (FIG. 15)

This type is identified by its black bullet. As with the ball cartridge, the cap annulus is black.

The cartridge case, cap and propellant charge are of the normal type.

The bullet consists of a steel envelope coated on both sides with gilding metal and containing a mild-steel core which is magnetic and is fitted with a lead cup behind the ogive. The weight of the bullet is 98·5 grains and that of the core is 53·5 grains. The pyramid diamond hardness figure for the core is 178.

The weight of the cartridge is approximately 166 grains.

INTERCHANGEABILITY

(i) 7·92-mm. German ammunition is interchangeable with British 7·92-mm. Besa ball.

(ii) 9-mm. German Pistol and Machine Carbine ammunition is interchangeable with British 9-mm. Parabellum.

(iii) 13-mm., 15-mm., 20-mm. and 30-mm. German ammunition is not interchangeable with British Service ammunition.

FIG. 1.
GERMAN 7.92 mm. CARTRIDGE. TYPICAL.

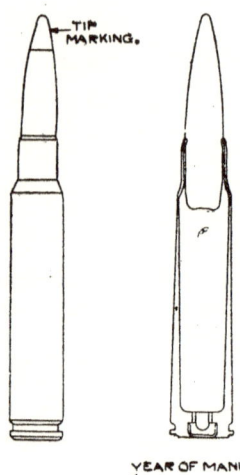

- TIP MARKING.
- DELIVERY No.
- STRIPE MARKING.
- YEAR OF MANUFACTURE.
- MANUFACTURE.
- S* BRASS CASE.
- S STEEL CASE.

FIG. 2.
GERMAN 7.92 mm. BALL. s.S.

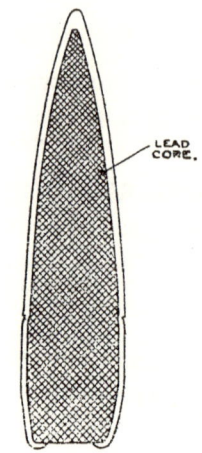

- LEAD CORE.

FIG. 3.
GERMAN 7.92 mm. ARMOUR PIERCING. S.m.K.

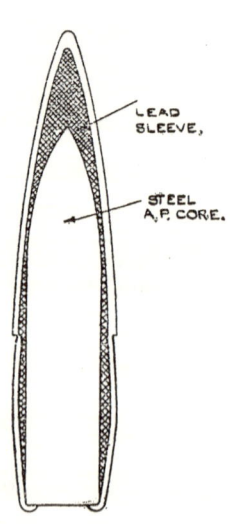

- LEAD SLEEVE.
- STEEL A.P. CORE.

FIG. 4.
GERMAN 7.92 mm. A.P./T. S.m.K. L'SPUR./

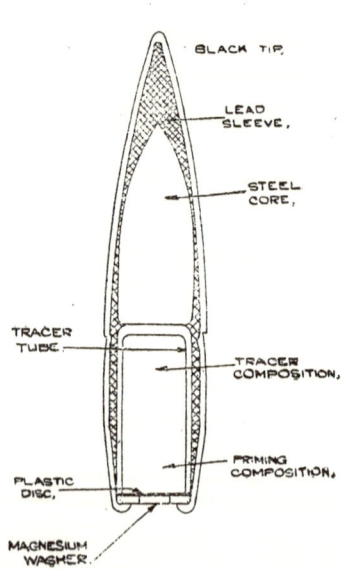

- BLACK TIP.
- LEAD SLEEVE.
- STEEL CORE.
- TRACER TUBE.
- TRACER COMPOSITION.
- PRIMING COMPOSITION.
- PLASTIC DISC.
- MAGNESIUM WASHER.

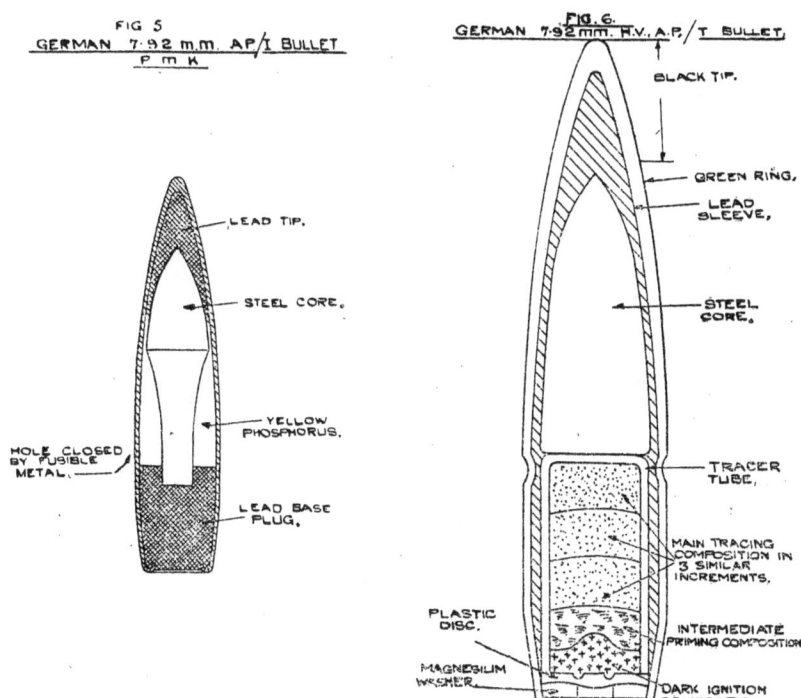

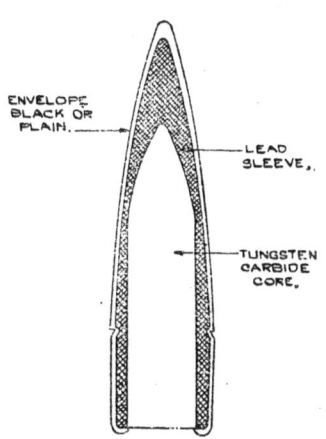

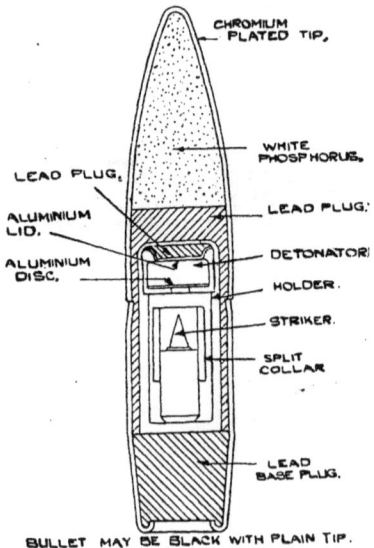

FIG. 8.
GERMAN 7·92 mm. EXPLOSIVE INCENDIARY BULLET.

BULLET MAY BE BLACK WITH PLAIN TIP.

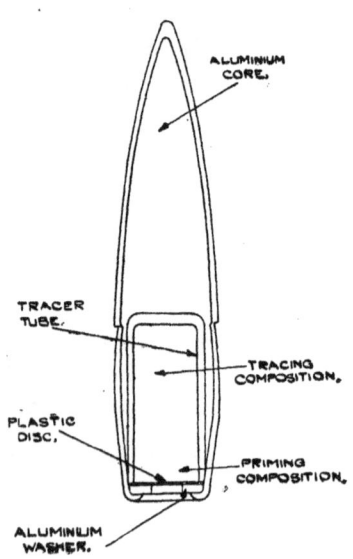

FIG. 9.
GERMAN 7·92 mm. PRACTICE, TRACER I.S.L SPUR.

FIG. 10.
GERMAN 7·92 mm. S.A.A. PACKAGE LABELS.

BALL CARTRIDGES.

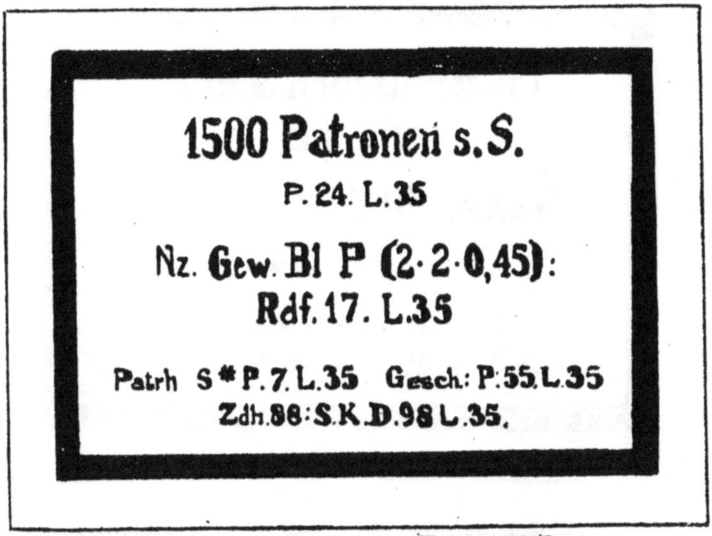

```
1500 Patronen s.S.
     P. 24. L. 35
Nz. Gew. Bl P (2·2·0,45):
     Rdf. 17. L. 35
Patrh S*P.7.L.35  Gesch: P.55.L.35
     Zdh.88: S.K.D.98 L.35.
```

PRACTICE BALL CARTRIDGES.

```
1500 Patronen l.S
       P.1.L.35
Nz. Gew. Bl.P. (2·2·0,45):
      RdL 5. L. 35
Patrh.: S* P.38.L.35  Gesch: P.1.L.35
     Zdh. 30: S.K.D. 139 L.35
```

COLOURED BAND.

FIG. II.
GERMAN 7·92 mm. S.A.A. PACKAGE LABELS.

ARMOUR PIERCING CARTRIDGES.

```
1500 Patronen S.m.K.
       P.2.L.35
Nz.Gew. Bl.P. (2·2·0,45):
    Rdf.
    126  1.L.35
Patrh: S*P.57.L.35   Gesch. P.77 L.35
GeschoBteile:P.   Zdh.88 S.K.D.58.L.35
```

ARMOUR PIERCING TRACER CARTRIDGES.

```
1500 Patronen S.m.K L'spur grünrot
           P.1.L.35.
Nz.Gew. Bl.P. (2·2·0,45) : Rdf.17.L.35.
Patrh: S*P.69.5.L.35   Gesch: P.69.20.L.35
    GeschoBteile:P.69   Satz:R.W.S.
         Zdh 88: S.K.D.20.L35

Trocken aufbewahren! Gegen StoB und Fall schützen!
```

FIG. 12.
GERMAN 7·92 mm. S.A.A. PACKAGE LABELS.

EXPLOSIVE INCENDIARY CARTRIDGES.

> **1500 B.-Patronen**
>
> P. 11 L. 35
> Nz.Gew.B 1.P.(2·2·0,45):
> Rdf. 5. L. 35
> Patrh· S*P. 39. L. 35. Gesch: P. 1. L. 35
> Zdh. 88; S. K. D. 430 L. 35.
> Gegen Stoß und Fall schützen!

PRACTICE TRACER CARTRIDGES.

> 1500 Patronen l.S L'spur (gelb)
> P. 1. L. 35.
>
> Nz.Gew. B 1.P (2· 2· 0,45) Rdf. 5. 35.
> Patrh.: S*P 39. L. 35. – Gesch: P. 1. L. 35.
> Satz: P. 1. L. 35 Zdh. 30; S. K. D. 139 L. 35.
>
> Trocken aufbewahren! Gegen Stoß und Fall schützen!

↘ COLOURED BAND.

FIG. 13.
GERMAN 9 M.M. BALL PISTOL CARTRIDGE.

FIG 15
GERMAN 9 mm. SEMI ARMOUR PIERCING.

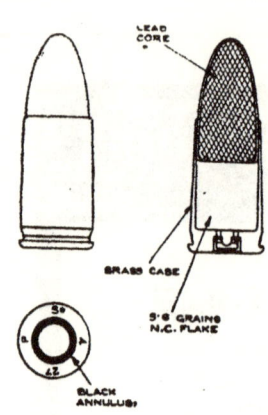
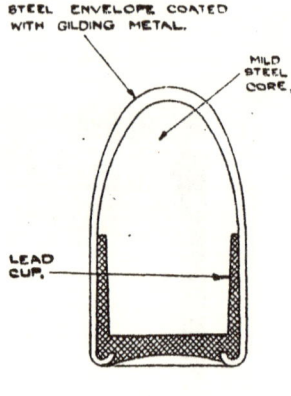

FIG. 14
GERMAN 9mm. PISTOL CARTRIDGE PACKAGE LABEL.

BALL CARTRIDGE.

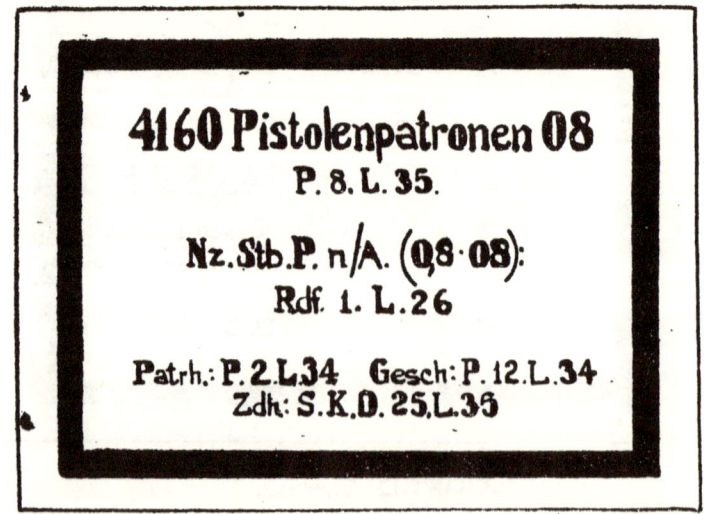

MARKINGS ON GERMAN AMMUNITION FOR 13-mm. TO 30-mm. WEAPONS

The following identification markings are used on the undermentioned ammunition :—

13-mm. Solothurn (·512-inch) (fired by 13-mm. aircraft gun MG 131).
15-mm. Mauser (·591-inch) (fired by 15-mm. aircraft gun MG 151).
20-mm. Oerlikon (·787-inch) (fired by 2-cm. aircraft gun MG 151).
20-mm. Solothurn (·787-inch) (fired by 2-cm. A.A./A. Tk. gun and 2-cm. tank gun).
20-mm. Mauser (·787-inch) (fired by 2-cm. aircraft gun MG 151).
30-mm. Solothurn (1·181-inch) (fired by 3-cm. aircraft gun).

Colour of Projectile

The body is painted to indicate the nature of the projectile. The colours used and their indications are :—

Yellow	H.E.
Black	A.P.
Olive Green	Ball.
Grey	Practice.

BANDS

Coloured bands on the projectile are used to indicate—(a) colour of trace, (b) absence of self-destroying arrangement, (c) certain natures of filling and (d) rounds for use in tropics.

(a) Colour of trace

With the exception of 20-mm. Oerlikon light case H.E. shell (marked with a black ring near the head) all H.E. shell provide a trace, but this is not indicated by a band if the colour of the trace is the normal pale green or white. The bands used for other colours and for ball or A.P. are adjacent to the driving band and vary in width from 2·5 mm. to 6 mm. according to the calibre. The colours of the bands and the corresponding traces are :—

Red band	Red trace.
White band	Brilliant white trace.
Yellow band	Bright pale green trace.

(b) Absence of self-destroying arrangement

None of the 13-mm. projectiles is self-destroying, but no special marking indicates this. 15-mm. and 20-mm. projectiles have been found marked immediately above the driving band or below the fuze. The marking used is :

Black band	Indicates, not self-destroying.

(c) Nature of filling

Yellow band. This marking is used near the ogive of the 30-mm. A.P. shell with an H.E. filling to indicate the H.E. filling.

Blue band. This marking round the centre of the shell indicates an incendiary filling in addition to the H.E. bursting charge. In the case of the 13-mm. H.E./Incendiary shell the band has been found below the fuze.

(d) *Rounds for use in tropics*

Red band. The band, 2 mm. wide, is found immediately in front of the driving band and may be superimposed on the tracer band. The smaller calibres, i.e. 15 mm., have this band painted round the junction of the mouth of the case and driving band.

Exceptions

The 20-mm. Solothurn A.P.T. has a phosphorus filling but is not marked with a blue band. The letters " Ph " are stencilled in white.

A 20-mm. Solothurn shell painted yellow and marked with a 5-mm. black band of thin circumferential lines above the driving band has recently appeared. The shell has an H.E. filling with an incendiary mixing and has a pale green trace. A self-destroying arrangement is included.

GERMAN 2-CM. CARTRIDGES (Fig. 16)

The three types, Oerlikon (air service), Mauser (air service) and Solothurn (land service), as shown in the drawing, can be identified by the cartridge cases. The Oerlikon and Mauser cases are similar but the Oerlikon is smaller in diameter, is without a pronounced taper and shoulder below the mouth and the forward side of the groove near the base is radiussed. The Solothurn case is longer than the other types and has a positioning band, or belt, formed in front of the groove near the base.

In the following details of the three types of ammunition small variations in weights may, in some instances, be within the tolerances permitted in manufacture and, in the case of propellant charges, may be the result of the adjustment of charge weight for a particular propellant lot. In a number of instances the identification markings on the projectiles do not conform with the principles of the system described in this pamphlet, but in most of these instances the projectiles are of the older types, many of which are not now in use. The colours of the traces are those produced when the composition is ignited with the projectile static. With the projectile in flight it is possible that these colours may appear different.

OERLIKON TYPES

Cases and Caps

The case is necked and is of the rimless type with the usual cap chamber with anvil and two flash channels formed in the base. Cases may be of brass or steel, the steel type being lacquered internally and externally.

The cap annulus with all types is black.

The brass cap contains a 2·35-grain charge covered by a foil disc and consisting of mercury fulminate, potassium chlorate, antimony sulphide and glass.

Propellant Charges

The propellant is in the form of graphited flakes of nitrocellulose powder and consists basically of approximately 95 per cent. of nitrocellulose (Nitrogen content 13·25 per cent.) with diphenylamine and diethyl diphenylurea. The approximate dimensions of the flakes are ·05 inch square and ·013 inch thick.

GERMAN 2-CM. OERLIKON CARTRIDGES

Nature and Fuze	Projectile markings			Projectile Weights			Propellant N.C. Flake	Trace Colour	Cartridge Wt.	Fig. No.
	Body	Bands		Filling	Total					
Ball ...	Olive green	None		None	grains 2053		grains 225·5	None	grains 3066	17
Tracer ...	Olive green	Yellow driving band with mauve above.		None	2053		225·5	White bright	3066	18
Tracer ...	Olive green	Yellow above driving band. White on nose with 3 arrows stamped in band		None	2094		225·5	Pale green.	3107	19
H.E. A.Z. 1502 or H.E. (S.D. Fuze)Z.Z.1505	Yellow	Black below fuze		262 grains P.E.T.N./Wax	1385		225·5	None	2398	—
H.E./T.(S.D.) A.Z. 1501.	Yellow	Black above driving band.		57 grains P.E.T.N./Wax.	2080		225·5	Yellow bright	3093	20
H.E./T.(S.D.) A.Z. 5045.	Unpainted steel.	White above driving band.		49 grains P.E.T.N./Wax.	2080		225·5	White bright	3093	—
H.E./T.(S.D.) A.Z. 1504	Yellow	None		57 grains P.E.T.N./Wax	1844		202	—	2834	—
H.E./I/T.(S.D.) A.Z. 1504	Yellow	Blue at centre of body or below fuze.		55 grains P.E.T.N./Wax 6 grains incendiary composition.	1742		225·5	Pale green	—	21
A.P. Shot ...	Black	None		72 grains Inert plastic.	1784		225	None	2773	22

Ball (FIG. 17)

The steel projectile has a flat-nosed ogival head, is fitted with a copper driving band and is cannelured behind the driving band for the attachment of the cartridge case. A cylindrical cavity formed in the shot is closed at the base by a plug which is driven in against a shoulder formed in the cavity wall.

Tracer Shell (FIGS. 18 and 19)

The steel projectile is fitted with a steel, flat-nosed ogival plug in place of a fuze. A cannelure is formed behind the copper driving band for the attachment of the cartridge case. Two cavities formed in the interior of the projectile are separated by a diaphragm formed in the body. The front cavity is empty. The rear cavity contains pressings of tracer and priming compositions and is closed at the base by a disc of aluminium foil held between two steel washers.

The following markings and traces have been reported :—

> Mauve and yellow bands above the driving band, white bright trace.
> Yellow and white bands above the driving band, white bright trace.
> Yellow band above driving band and white band on nose plug with three vertical arrows stamped in white band, bright white trace.
> Yellow band above driving band, pale green trace.

H.E. Shell

A drawing and description of this light-case shell and the fuze A.Z. 1502, are included in Pamphlet No. 3.

H.E., Self-destroying, Shell

This is the light-case H.E. shell fitted with the self-destroying fuze Z.Z. 1505 instead of the A.Z. 1502. A drawing and description of the Z.Z. 1505 are included in this pamphlet.

H.E., Tracer, Self-destroying, Shell with Fuze A.Z. 1501 (FIG. 20)

The mild-steel shell is prepared to receive a nose fuze, is fitted with a copper driving band and is cannelured behind the driving band for the attachment of the cartridge case. The interior is designed to form two compartments separated by a diaphragm forming part of the body. The diaphragm has a central hole. The front compartment contains a bursting charge consisting of penthrite/wax (approximately 85/15) in the form of a pellet with a cavity to receive the magazine of the fuze at the front end and a cavity to receive the self-destroying unit at the base end. The rear compartment contains the tracing and priming composition and is closed by a disc of aluminium foil held between two steel washers. The self-destroying pellet of H.E. in a metal container is fitted into the hole in the diaphragm from the front and thus leads from the tracing composition to the bursting charge.

The self-destroying H.E. pellet consists of potassium nitrate, potassium picrate and carbon.

The tracer composition produces a bright white trace and consists of boiled oil, barium nitrate, magnesium and hammerscale.

The priming composition consists of hammerscale, barium peroxide and magnesium.

H.E., Tracer, Self-destroying, Shell with Fuze A.Z. 5045

This shell is similar to the yellow, black-banded, shell of the same nature (Fig. 20), but has a larger tracer cavity and a smaller cavity for the bursting charge.

H.E. Tracer, Self-destroying, Shell with Fuze A.Z. 1504

This shell is the same as the yellow, black-banded, shell of the same nature (Fig. 20), but is fitted with the fuze A.Z. 1504 instead of the A.Z 1501.

H.E., Incendiary, Tracer, Self-destroying, Shell (FIG. 21)

This shell is the same as the H.E./T/S.D. shell but has an incendiary pellet in a cavity formed in the H.E. filling below the cavity accommodating the magazine of the fuze. The fuze A.Z. 1504 is used instead of the A.Z. 1501. A drawing and description of the A.Z. 1504 are included in this pamphlet.

The incendiary pellet is secured in the cavity in the H.E. pellet by varnish and consists of ferric oxide, magnesium, aluminium, zirconium and grease.

The self-destroying pellet, in a brass container, consists of potassium dinitrophenate, potassium nitrate and charcoal. The priming composition in the tracer consists of barium peroxide and magnesium with an organic binding material.

The tracing composition consists of barium nitrate, magnesium, sodium borate and resin.

Armour Piercing Shot (FIG. 22)

The steel pointed projectile is fitted with a sintered iron driving band and is cannelured for the attachment of the cartridge case. The cavity formed in the interior, and containing an inert filling similar to bakelite, is large for a piercing projectile and is probably designed to reduce the weight. The projectile is closed at the base by a light alloy plug. The same projectile is used in the Mauser cartridge as an A.P. shell filled H.E. and as a A.P./I. shell charged phosphorus.

With normal impact at 100 yards the projectile will perforate a 17-mm. homogeneous hard plate. With impact at 30 degrees to the normal at the same range the projectile will defeat a 10-mm. plate of the same type.

MAUSER TYPES

Cases, Caps and Electric Primers

The steel case is necked and of the rimless type. The interior and exterior are treated with lacquer or plated. The case is about the same length as the Oerlikon type but has a larger capacity and differs in shape. The usual cap chamber with anvil and two fire channels is formed in the base. The cap annulus is normally black, but cases fitted with electric primers have been found to have red or green colouring.

The cap is of brass and contains a 1·56-grain charge of composition which is varnished on the surface and covered by a metal foil. The foil disc is turned up at the edges and adheres to the wall of the cap. The coper surface of the foil is treated with nitrocellulose lacquer tinted green. the composition consists of potassium chlorate, antimony sulphide, calcium silicide and tetrazene.

Electric primers have been found in H.E./I. (S.D.) and H.E./I./T.(S.D.)

GERMAN 2-CM. MAUSER CARTRIDGES

Nature and Fuze	Projectile Markings		Projectile Weight		Propellant N.C.T.	Trace Colour	Cartridge Wt.	Fig. No.
	Body	Bands	Filling	Total				
H.E. A.Z. 1502	Yellow	Black below fuze	262 grains P.E.T.N./Wax	grains 1385	grains 308	—	grains 2671	—
H.E.(S.D.) Z.Z. 1505	Yellow	Black below fuze	262 grains. P.E.T.N./Wax	1333	310	—	2700	—
H.E./1/T.(S.D.) A.Z. 1504	Yellow	·2-inch blue ·4 inches above driving band or ·25-inch blue below fuze.	55 grains P.E.T.N./Wax 6 grains incendiary pellet.	1742	281	Pale green	3080	21
A.P. Shot ...	Black	None	72 grains inert plastic (or liquid)	1784	283	—	3190	22
A.P./H.E. ...	Black	·7-inch yellow ·2 inches above driving band.	78 grains P.E.T.N./Wax	1794	291	—	3126	24
A.P./I. ...	Black	·2-inch Blue ·7 inches from nose.	52 grains phosphorus	1802	290	—	3140	25
H.E./1 (S.D.) Z.Z. 1505 with delay.	Yellow	Black below fuze.	271 grains	1420	N.C. Flake 301 grains	—	2772	—

cartridges. In these instances the annuli at the base of the cases were red and green respectively. The cap chamber is enlarged to take the primer which is secured by stabbing at three points. The primer, as shown in Fig. 23, consists of a brass cylindrical body containing a brass contact plug which is separated from the body by an insulating cup and has an exposed projecting piece at the base. A fuze head consisting of two contact strips, separated by a strip of insulating material, and connected at one end by a bridge wire embedded in a blob of ignition composition, is positioned within the insulating cup with one strip in contact with the plug and the other in contact with a brass contact washer fitted above the cup. The washer is in contact with the body and is provided with a projecting piece which bears on the upper contact strip of the fuze head. A brass collar, forming a surround to the magazine filling, retains the contact washer and supports a thin brass closing disc over which the body is turned.

The path of the current is through the contact plug to the lower contact strip of the fuze head, through the wire bridge to the upper contact strip and thence by the contact washer to the body of the primer. The heating of the bridge wire ignites the fuze head composition and results in the firing of the magazine filling.

Propellant Charges

The propellant consists of irregular tubular grains of graphited nitrocellulose powder. The approximate dimensions of the grains, in inches, are :- length ·059, external diameter ·035, internal diameter ·005. The composition consists basically of approximately 95 per cent. of nitrocellulose (nitrogen content approximately 13 per cent.) stabilized with diphenylamine and contains ethyl centralite and potassium sulphate.

H.E. Shell

The light-case, hemispherical-based shell, fitted with the direct action fuze No. 1502 is the same as that used in the Oerlikon type. The shell and fuze are described in Pamphlet No. 3.

H.E. Shell with Self-destroying Fuze Z.Z. 1505

This is the light-case, hemispherical-based shell as used in the Oerlikon type (described in Pamphlet No. 3), but the self-destroying fuze Z.Z. 1505 is used instead of the A.Z. 1502. A drawing and description of the Z.Z. 1505 are included in this pamphlet.

H.E. Incendiary Shell with Self-destroying Delay Fuze Z.Z. 1505

This is the light-case, hemispherical-based shell as used in the Oerlikon type (described in Pamphlet No. 3), but has a cast filling consisting of an H.E. intermixed with a metallic substance.

Cartridges of this type have been found with the delay magazine in the Z.Z. 1505 fuze and an electric primer instead of a percussion cap in the case.

Drawings and descriptions of the delay magazine of Z.Z. 1505 and the electric primer are included in this pamphlet.

H.E. Incendiary, Tracer, Self-destroying, Shell (FIG. 21)

This shell is the same as that described for the Oerlikon type.

Cartridges of this nature have been found with percussion caps and N.C. tubular propellant or with electric primers and N.C. flake propellant.

Armour Piercing Shot

This projectile with its inert filling is the same as that used in the Oerlikon type and is shown in Fig. 22.

Armour Piercing Shell, Filled H.E. (FIG. 24.)

The shell body is the same as that of the A.P. shot shown in Fig. 22, but the cavity contains a filling of P.E.T.N./Wax. A gaine is fitted in a cavity formed in the base of the filling and the base of the shell is closed by an igniferous base fuze. A cavity formed in the fuze body contains an aluminium cylinder fitted with a small initiator pellet of the igniferous type at the front end and open at the base end. A needle, with two splayed projecting pieces, is positioned at the open base end and is prevented from approaching the pellet by the two projecting pieces bearing against the wall of the cylinder. The fuze is closed at the front end by a screwed plug with a small central hole. The gaine consists of an aluminium flanged cup inserted in the cavity in the bursting charge with its open end towards the base. The cup contains a filling of the intermediary class which is topped near the open end by an initiator composition. An empty space is left at the open end of the gaine.

On impact the needle overcomes the projecting pieces and, moving forward, pierces the igniferous pellet. The flash produced passes through the small hole and initiates the gaine thus bringing about the detonaton of the bursting charge.

Armour Piercing Incendiary Shell (FIG. 25)

The shell body is the same as that of the A.P. shot shown in Fig. 22 but, instead of an inert filling, the cavity contains a light alloy container charged with phosphorus.

SOLOTHURN TYPES

Cases and Caps

The case is necked and rimless and has a belt of high diameter formed in front of the groove near the base. The usual cap chamber with anvil and two fire channels is formed in the base. Cases may be of brass or of steel. The steel cases are lacquered internally and externally. Brass cases are attached to the projectile by indenting or rolling of the neck into the cannelure formed in the projectile. Steel cases are not secured in this way and appear to rely upon being a close fit. The cap annulus, with all types, is black. Steel cases may also be coppered or gilded.

The cap is of brass and is not secured by ringing or stabbing. The cap composition is the same as that used in the 7·92-mm. cartridge. The composition is varnished on the surface and covered by a foil disc which is kept in position by a thin brass sleeve fitting tightly into the wall of the cap. The weight of the composition is approximately ·7 grains.

Propellant Charges

The propellant consists of irregular tubular grains of nitrocellulose powder, which are graphited. The approximate dimensions of the grains, in inches, are : length ·08, external diameter ·054, internal diameter ·01. The composition consists basically of approximately 93 per cent. of nitrocellulose (nitrogen content 13·27 per cent.) stabilized with diphenylamine and moderated with centralite. Potassium sulphate is also included and in some instances camphor.

GERMAN 2-CM. SOLOTHURN CARTRIDGE (fired by 2-cm. A.A./A. Tk. gun and 2-cm. tank gun)

Nature and Fuze	Projectile Markings		Projectile Weights			Propellant N.C.T.	Trace colour	Cartridge Wt.	Fig. No.
	Body	Bands	Filling	Total					
H.E./T. (S.D.) A.Z. 5045	Yellow	—	96 grains P.E.T.N./Wax	grains 1775		grains 618	Pale green	grains 4676	26
H.E./T(S.D.)S/L A.Z. 5045	Yellow	·08-inch red, above driving band.*	94 grains P.E.T.N./Wax	1840		564	Pale green	4676	27
H.E./I/T (S.D.) S./L. A.Z. 5045	Yellow	·24-inch black ·08 inches above driving band.	105 grains P.E.T.N./Wax and Aluminium flake.	1837		564	Pale green	4480	—
A.P./I./T.	Black Stencilled "Ph" in white.	·2-inch yellow above driving band.	42 grains phosphorus.	2260		564	Pale green	4946	28
A.P./T. (Filled Sulphur)	Black Stencilled "O" in white.	·08-inch red* above driving band and ·2-inch yellow above.	55 grains Sulphur and Sand.	2260		564	Pale green	4946	29
A.P./T. (Filled Sulphur)	Black with ·4-inch red tip.	·08-inch red above driving band.*	52 grains Sulphur and Sand.	2265		551	Green to red	4960	—

* The red band indicates a propellant charge weight for hot climates and may be found on all types of projectile.

The charge is contained in a silk bag, an igniter being formed at the base of the bag. The igniter consists of a small pocket containing approximately 15·8 grains of gunpowder.

H.E., Tracer, Self-destroying Shell (FIG. 26) (2-cm. Sprgr. (Patr.) L'Spur)

The shell is fitted with a copper driving band and is cannelured behind the band. The mouth of the case is not indented into the cannelure. Two cavities formed in the shell are separated by a diaphragm, with a central hole, formed in the shell body. A self-destroying pellet in a brass container is fitted in the hole in the diaphragm from the front.

The forward cavity contains the bursting charge in the form of a 94-grain P.E.T.N./Wax pellet. The pellet is recessed at the front end to receive the magazine of the fuze and also at the base end to fit over the self-destroying pellet.

The rear cavity in the shell contains a steel tube into which is pressed the tracing and priming compositions. The tube is secured in position by a steel hollow base plug, the base plug being closed by celluloid disc. The trace is pale green and burns statically for 6·4 seconds.

The fuze A.Z. 5045, used in this shell, is described in this pamphlet.

H.E., Tracer, Self-destroying, Streamlined Shell (FIG. 27) (2-cm. Sprgr. (Patr.) L'Spur)

The streamlined shell body is similar to that of the H.E./T.(S.D.) plain-base shell, but is heavier. The self-destroying pellet is not enclosed in a metal container and appears to be put in during the moulding of the H.E. bursting charge. The H.E. bursting charge of P.E.T.N./Wax is of the same weight as that in the plain-base shell, i.e. 94 grains.

The tracing and priming compositions are pressed direct into the cavity in the rear part of the shell. A domed steel plug, coated with gilding metal is fitted in the forward end of the tracer cavity. The domed portion of this plug contains an ignition charge which is caused to flash and ignite the self-destroying pellet by the rise in temperature as the tracing composition burns away. The trace is pale green and burns statically for 6·4 seconds.

The fuze A.Z. 5045, used in this shell is described in this pamphlet.

H.E., Incendiary, Tracer, Self-destroying, Streamlined Shell (2-cm. Sprgr. (Patr.) L'Spur)

The shell is similar to the H.E./T.(S.D.) streamlined type (Fig. 27), but has a driving band of sintered iron. The 3·7-grain self-destroying pellet consists of irregular grains of ungraphited chopped propellant and is pressed into the cavity in the base of the high explosive/incendiary bursting charge pellet. The bursting charge pellet weighs approximately 105 grains and consists of 71 per cent. P.E.T.N./Wax and 29 per cent. of aluminium. The aluminium is mixed in with the H.E. in the form of irregular flakes of thin foil, ranging in size from 5×4 mm. (maximum) to $1·4 \times ·5$ mm. (minimum). The pellet is held lightly in compression in the shell by paper washers under the fuze.

Armour Piercing, Incendiary, Tracer Shell (FIG. 28) (2-cm. Pzgr. (Patr.) L'Spur)

The steel shell is fitted with a copper driving band and is cannelured presumably for the attachment of the cartridge case, although rounds examined have not had the case indented into the cannelure. The pyramid diamond hardness figure for the point of the shell is 982. This figure decreases gradually to 527 at the base. The cavity in the interior is of the small type characteristic of piercing shell and contains 46 grains of yellow phosphorus. The charging is sealed by a lead disc supported by a steel disc, both being secured in position by the solid end of the steel tracer plug. The tracer plug is in the form of a cup, screw-threaded externally to screw into the shell and lined internally with a copper-plated steel tube which contains the pressed tracing and priming compositions. The base end of the tube is closed by a celluloid disc secured with a varnish.

The priming composition consists of barium peroxide and magnesium with a binding material. The tracing composition includes barium nitrate, magnesium, sodium oxalate, French chalk and resinous matter. The trace is pale green and burns statically for 2·4 seconds.

Armour Piercing (Sulphur) Tracer Shell (FIG. 29) (2-cm. Pzgr. (Patr.) L'Spur)

This projectile is identical with the A.P./I./T. shell except that the cavity contains a 55-grain filling consisting of sulphur and fine sand. The tracer fitment is as described for the A.P./I./T. shell.

At 100 yards and normal impact, the projectile will penetrate with certainty up to at least 40-mm. armour plate, it may penetrate 50-mm. plate.

The sulphur is probably a substitute for the phosphorus or other charging.

Armour Piercing (Sulphur), Tracer Shell with Red Tip (2-cm. Pzgr. (Patr.) L'Spur)

This red-tipped shell (*see* table of markings) is similar to the previous type but has three steel discs inserted between the tracer plug and the sulphur pellet. The pellet weighs 52 grains and consists of 70 per cent. of sulphur and 30 per cent. of sand.

The trace is bright and changes from green to red.

GERMAN FUZE A.Z. 1501 (FIG. 30)

The fuze is of the detonating type with a direct action of the floating needle type and is used in 2-cm. shell.

The head of the fuze which screws on to the body is recessed at the front end to accommodate the hammer. The recess has a central hole through which the stem of the hammer passes and is closed against air pressure at the front by a metal disc. The hammer consists of a disc with a stem which is aligned with the needle in the fuze body.

The body is screw-threaded externally behind the flange for insertion in the shell and has an internal left-hand screw-thread for the magazine. A channel is formed through the centre of the body for the needle, the channel being of larger diameter at the front end to receive the head of the needle on functioning. A centrifugal safety slide, shaped to fit the stem of the needle below the enlarged head, is retained in the safe position in its slot by a detent carried in the body. The detent is contained in a sleeve which screws into a recess in the fuze body and is supported in contact with a chamfered portion of the safety slide by a pellet of gunpowder at its base. A flash channel in the recess behind the powder pellet leads to another recess containing a detonator held off a striker in the base of this recess by a spiral spring. The detonator contains a mixture of potassium chlorate and antimony sulphide topped with black powder. A similar empty recess is formed diametrically opposite in the fuze body; this is probably intended to preserve the balance of the fuze and to divert some of the pressure produced by the functioning of the detonator in the connected recess. Both recesses are closed at the front end by metal discs.

The magazine contains approximately 7 grains of P.E.T.N. in powder form and is closed by a detonator pressed in at the front end. The detonator contains about 15 grains of a mixture consisting of lead azide and calcium silicide topped with P.E.T.N.

Action

On acceleration the detonator in the recess sets back against its spring and is pierced by the striker. The flash passes through the channel to the detent recess and ignites the gunpowder pellet thus depriving the detent of this support. Centrifugal force, set up by the rotation of the projectile and fuze, causes the slide to move outwards, away from the needle head, the chamfered end of the slide depressing the detent as this movement occurs. During flight " creep " action and the protection from air pressure provided by the metal closing disc at the nose results in a forward movement of the hammer and needle. On impact the needle is driven into the detonator by the hammer and thus brings about the detonation of the magazine.

GERMAN FUZE A.Z. 1504 (FIG. 31)

The fuze is of the detonating type with a direct action and is used in 2-cm. and 3-cm. shell.

The head of the fuze is recessed at the nose to accommodate the aluminium hammer. The recess is closed against air pressure by a disc of brass foil. The hammer is in the form of a disc with a central stem which passes through a hole in the base of the recess and is aligned with the striker when the striker is in the armed position.

The aluminium striker is held in a carrier of the same material and, before arming, is displaced from the centre of the fuze body and the hammer. The carrier is recessed on the under side to receive the steel ball during flight.

The body is screw-threaded externally for assembly in the shell and internally to receive the magazine. Two holes to accommodate the centrifugal safety bolts are formed near its front end and a groove is formed around its exterior at this point to receive a copper retaining spring which retains the bolts in engagement with the striker carrier. A channel formed in a position displaced from the centre of the body contains the steel ball and has an extension formed along its length in which the striker is held by the two safety bolts before the fuze becomes armed. An aluminium washer, secured by stabbing, retains the ball in the channel.

The brass magazine is similar to that described for the A.Z. 1501.

Action

On acceleration the ball is held in the base of its channel by set back. During flight the ball moves forward into the recess in the carrier as the result of deceleration and thus gives a weight bias to the recessed side of the striker carrier. Centrifugal force set up by the rotation of the projectile causes the retaining spring to expand and the safety bolts to move outwards thus freeing the striker carrier. The ball weighted end of the carrier is then moved outwards and the striker is brought into alignment with the hammer. During flight the hammer is protected from air pressure by the brass-foil closing disc and is retained in the forward position by creep. On impact the hammer is forced in and drives the striker into the detonator.

GERMAN SELF-DESTROYING FUZE Z.Z. 1505 (FIG. 32)

The fuze is of the detonating type with a direct action and a self-destroying action which depends upon the loss of rotational velocity for operation. The fuze is used in 2-cm. shell.

The steel head of the fuze is plated with brass and has a large internal recess which houses the striker head. The recess is closed against air pressure at the nose by a metal closing disc.

The striker head is of light alloy and is fitted with a steel needle. A flange formed near the base of the striker head supports one end of a strong steel spiral spring and is grooved to carry eight steel balls. The balls are retained in position by a steel ring which increases in its internal diameter towards the front and is supported by the body of the fuze. The striker is supported by a split collar assembled round the needle and retained by a coil of strip brass. The split collar is supported by the body.

The steel body is plated with brass and is screw-threaded externally for the assembly of the head and for insertion in the shell. A central hole is formed for the needle and a brass magazine of the type described for the A.Z. 1501 or one of the delay type, shown in Fig. 33, is screwed in at the base.

In the delay type of magazine the intermediary filling and the detonator do not extend so far up the body of the magazine and the space thus provided in the upper portion is occupied by the delay fitment. The fitment consists of a light alloy plug, flanged at the head and of a suitable diameter to fit into the magazine. A small detonator of the igniferous type is fitted in the head of the plug, below the striker needle, and a flash channel behind the detonator communicates, by way of a groove round the body and another flash channel near the base, with the main detonator in the base portion of the magazine.

Action

During flight the steel balls in the striker head are moved outwards by centrifugal force against the inclined portion of the steel ring and in this position support the striker against the pressure of its spiral spring. The coil of the brass strip surrounding the split collar is also loosened and the split collar thrown clear of the striker. On impact the striker is forced in so that the inclined surface of the steel ring directs the balls back into their housing. With impact on a light plate the striker is then driven through the steel ring by its spiral spring and the needle pierces the detonator. Impact with a plate offering more resistance would probably result in the striker being driven directly into the detonator by the impact and consequently a more rapid action.

When impact does not occur before the rotational velocity has fallen sufficiently, the pressure of the striker spring overcomes the decreasing centrifugal force and causes the striker to move in. This movement directs the balls back into their housing thus permitting the spiral spring to drive the striker through the steel ring and the needle to pierce the detonator.

With the delay type of magazine the flash from the pierced detonator passes to the main detonator by way of the flash channels and the groove round the delay plug and in consequence causes a slight delay.

GERMAN FUZE A.Z. 5045 (Fig. 34)

The fuze is of the detonating type with a direct action of the floating needle type. The fuze is used in 2-cm. shell.

The aluminium head is recessed at the nose to accommodate a wooden hammer. The hammer is in the form of a disc with a stem on its underside which passes through a hole in the base of the recess and is aligned with the head of the striker. The recess is closed against air pressure by a brass closing disc at the nose. The head is screwed to the body and secured by a small set screw.

The aluminium body is screw-threaded externally for insertion in the shell and internally at the base to receive the magazine. The front face of the body is recessed to house the split collar and is bored centrally to receive the striker.

The striker assembly consists of a steel striker fitted with an aluminium head. The underside of the head is shaped to retain the brass split collar in the safe position. The split collar prevents the striker being driven into the detonator before the fuze is armed and is further retained by $1\frac{1}{2}$ turns of phosphor bronze strip surrounding it.

The magazine is of aluminium and is similar to the type described for the A.Z. 1501.

Action

On acceleration the split collar is held by the set back of the striker.

During flight the phosphor bronze spring uncoils and the split collar is thrown clear by centrifugal force. The striker, now unsupported, and the hammer are moved forward by creep action during deceleration and are protected from air pressure by the brass closing disc. On impact the hammer is forced in and drives the striker into the detonator in the magazine.

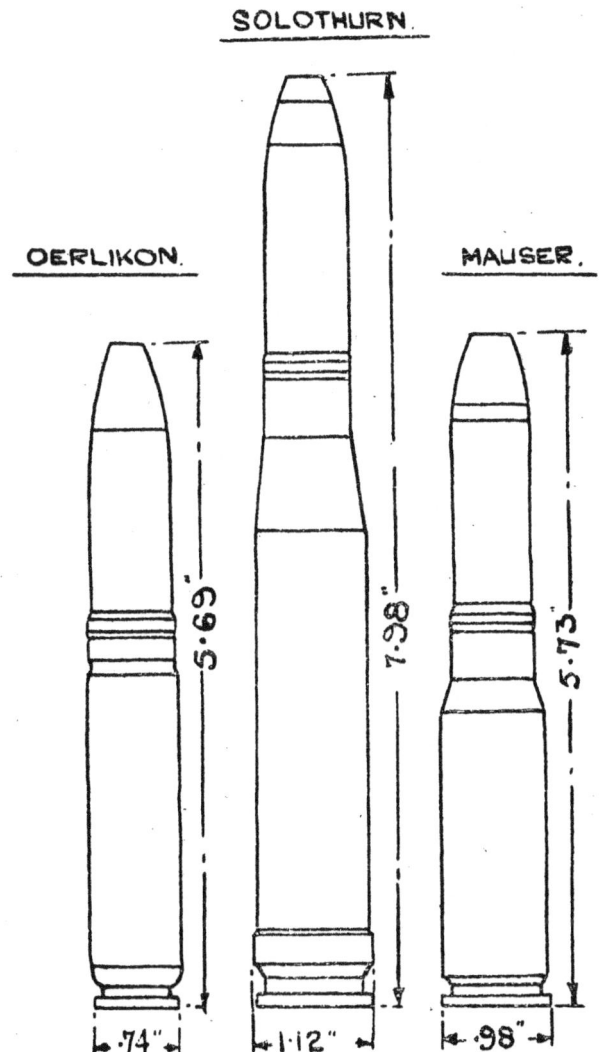

FIG. 16.
GERMAN 2 cm. CARTRIDGES.

FIG. 17.
GERMAN, 2 CM. OERLIKON BALL

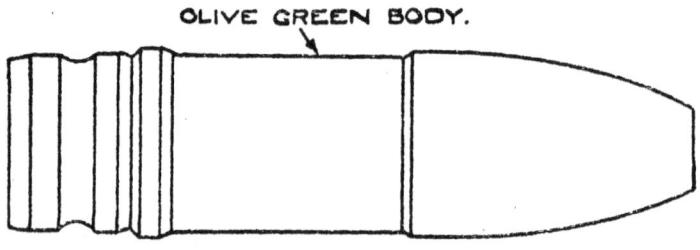

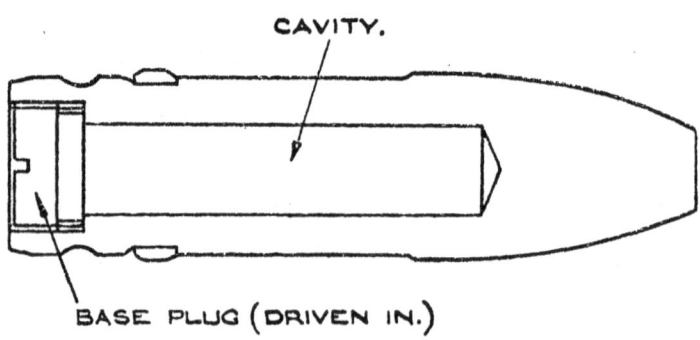

FIG. 18
GERMAN, 2 CM, OERLIKON, TRACER SHELL.

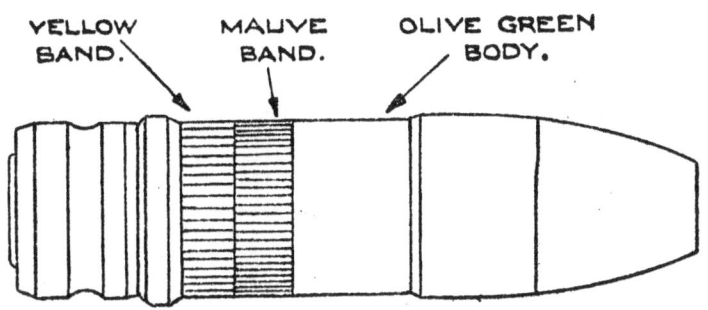

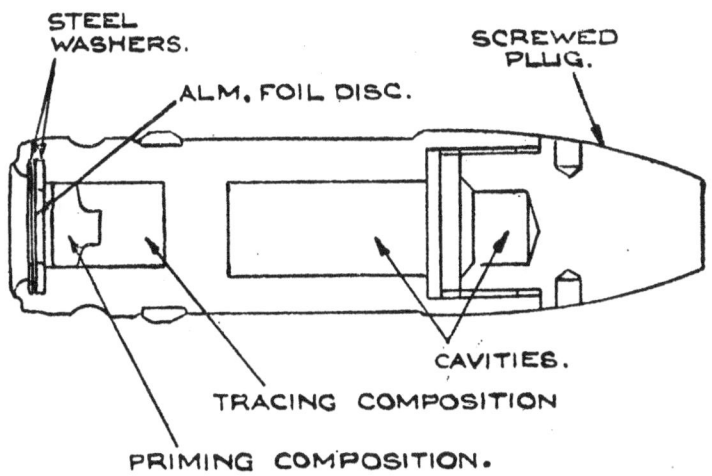

FIG. 19.
GERMAN 2 CM. OERLIKON TRACER SHELL.

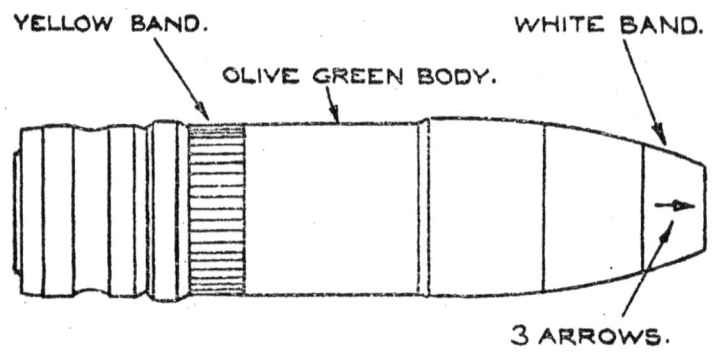

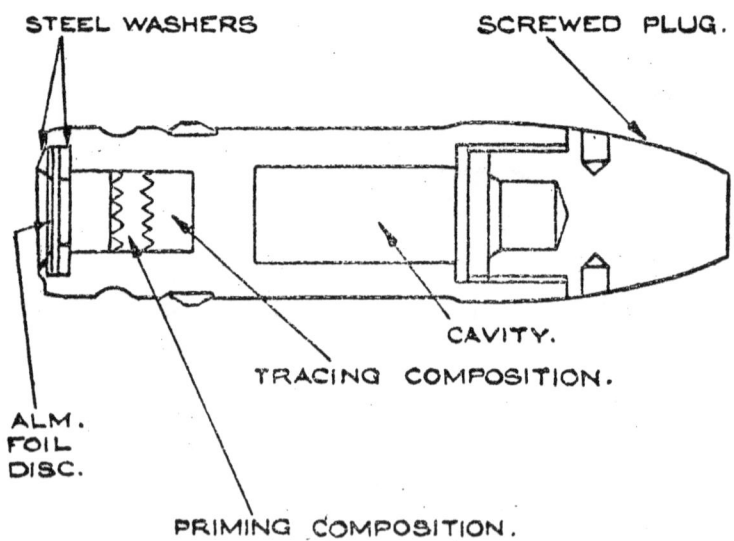

FIG. 20
GERMAN, 2 CM. OERLIKON H.E./T (S.D.) SHELL.

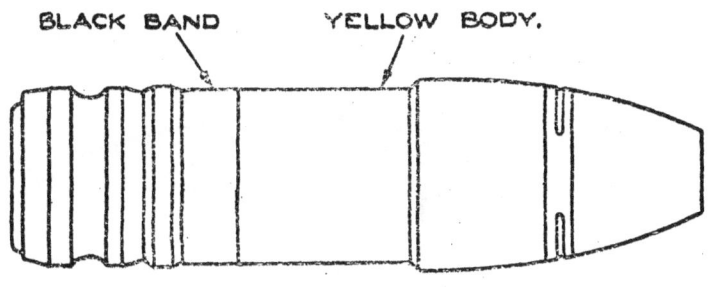

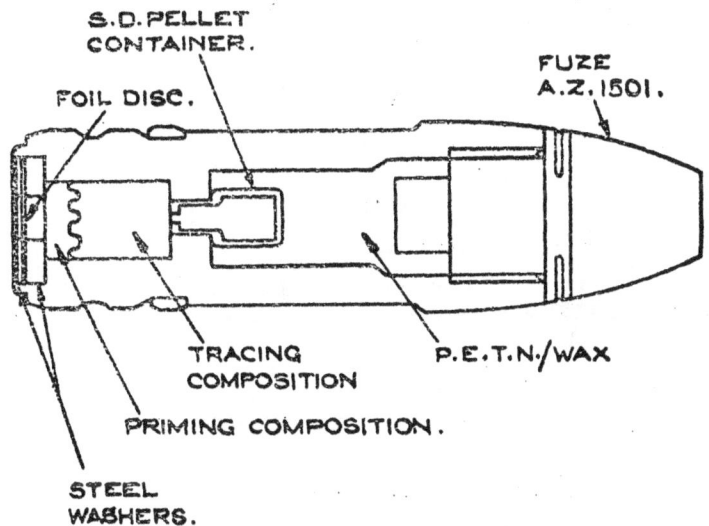

FIG. 21.
GERMAN, 2 CM, OERLIKON, H.E./I/T.(S.D.) SHELL.

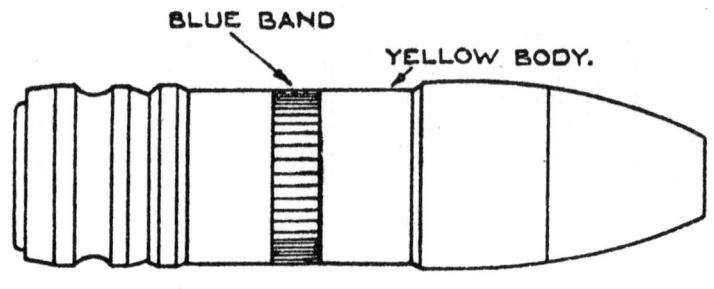

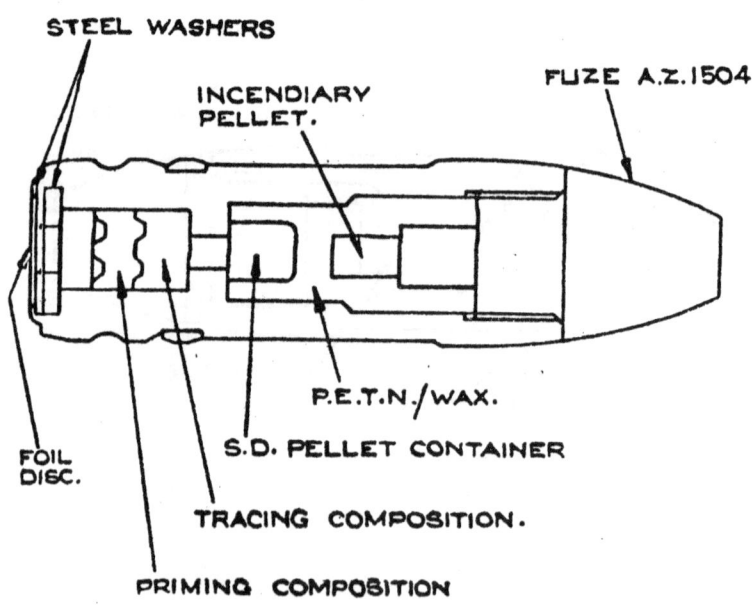

FIG. 22.
GERMAN, 2 CM, OERLIKON, A.P. SHOT.

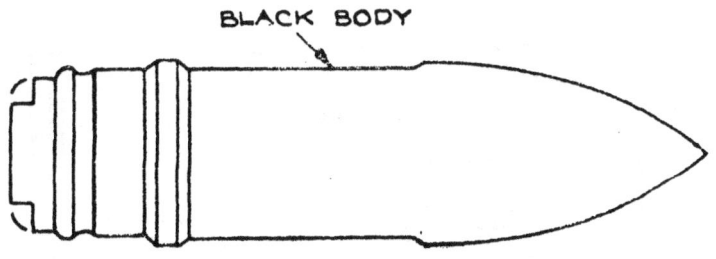

FIG. 23.
GERMAN ELECTRIC PRIMER FOR 2 cm. MAUSER CARTRIDGE.

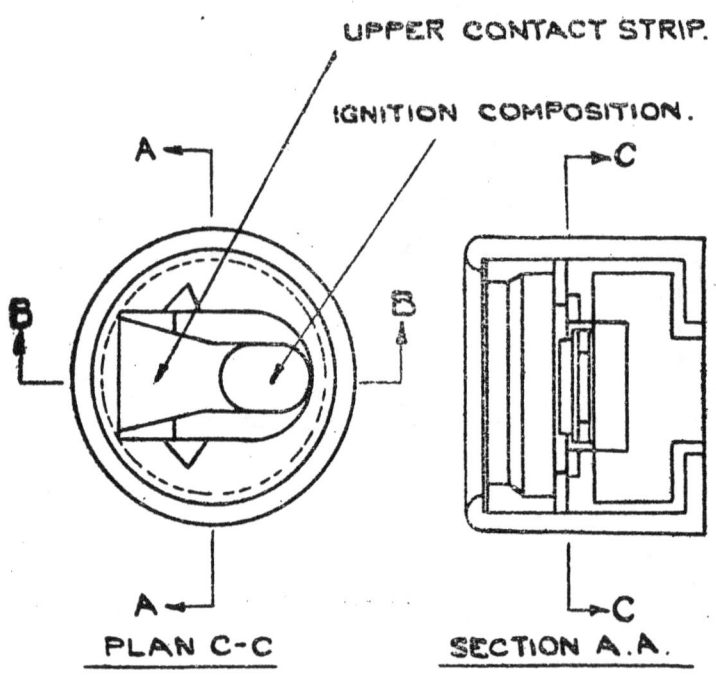

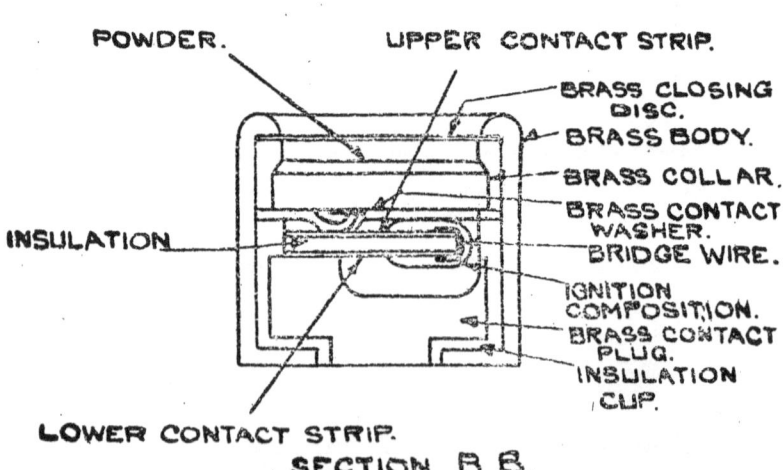

FIG 24
GERMAN 2 CM MAUSER A.P. SHELL FILLED H.E.

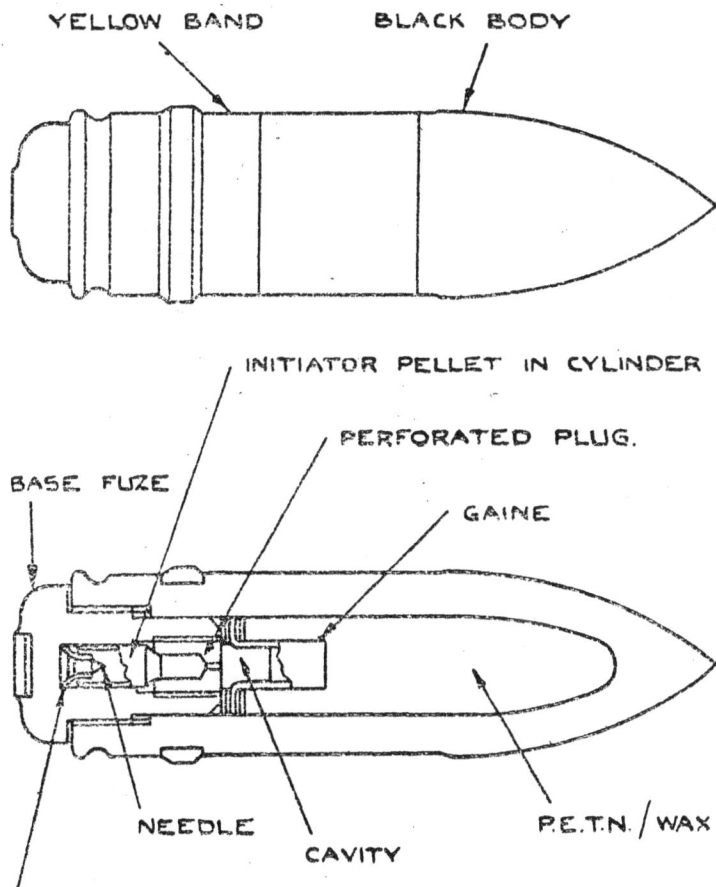

FIG 25
GERMAN 2 CM. MAUSER A.P./I SHELL

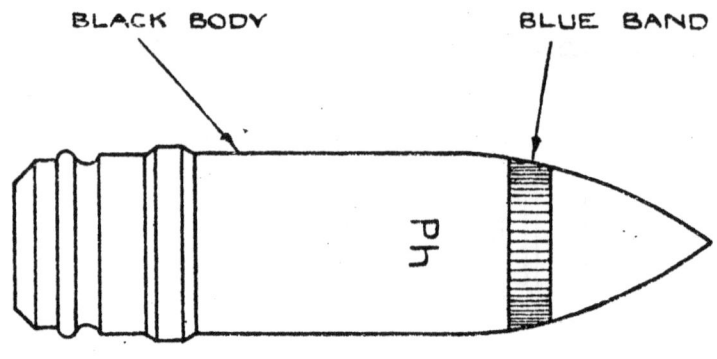

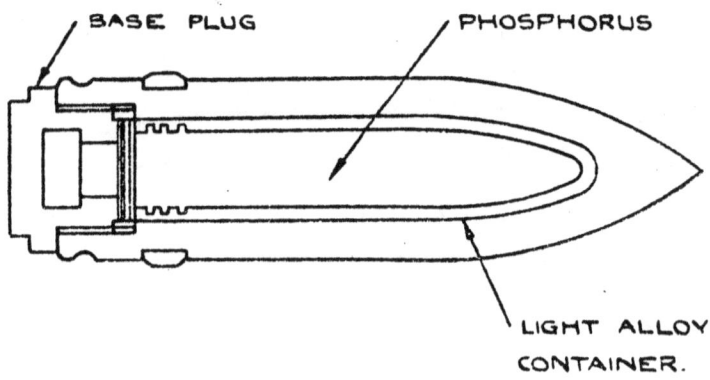

FIG 26

GERMAN 2 CM. SOLOTHURN H.E/T (S.D) SHELL

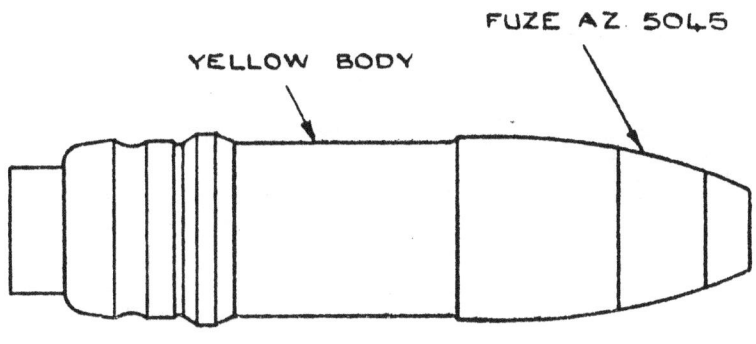

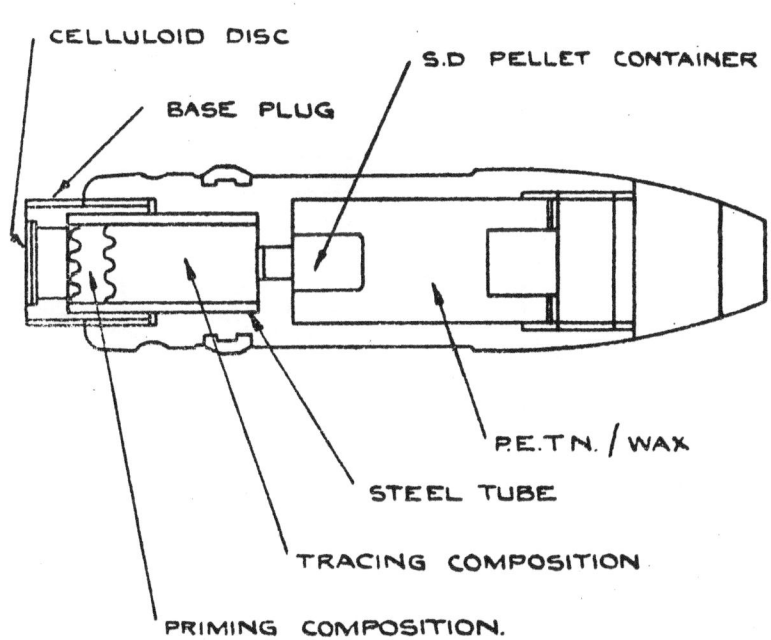

FIG 27

GERMAN 2 CM, SOLOTHURN, H.E./T. (SD) STREAMLINED SHELL.

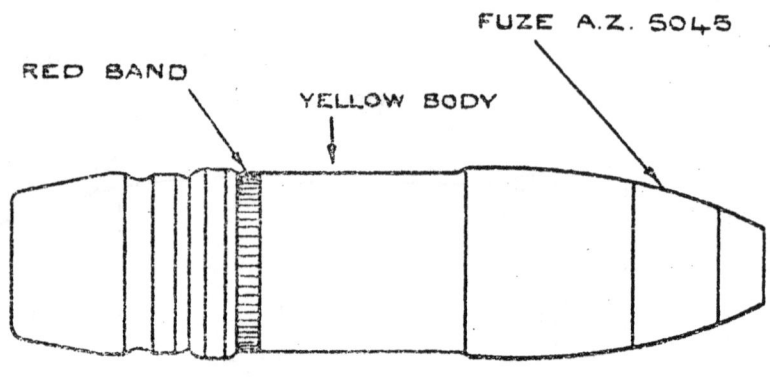

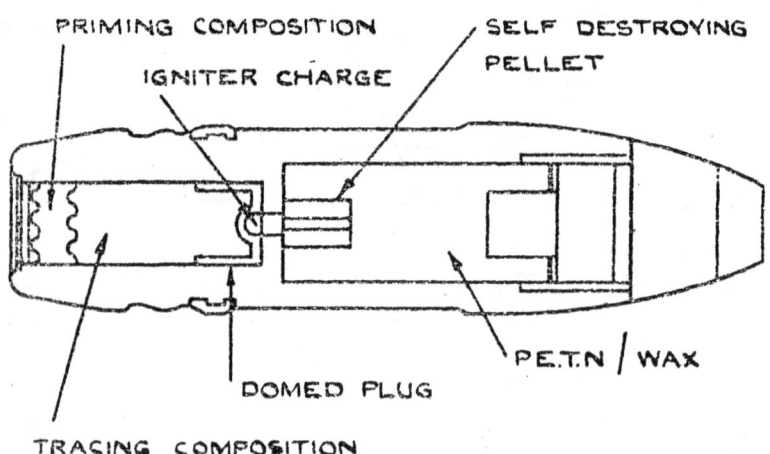

FIG. 28.
GERMAN 2 CM, SOLOTHURN, A.P./I./T. SHELL.

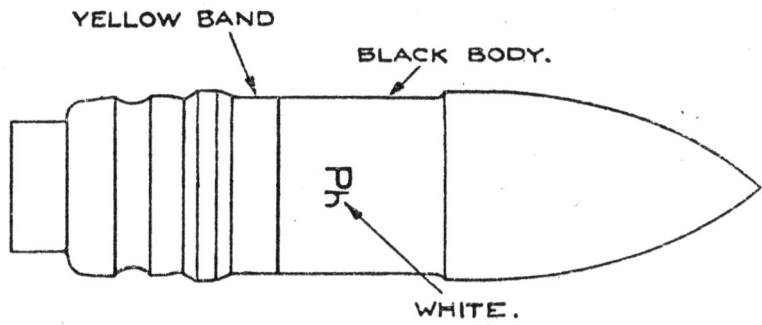

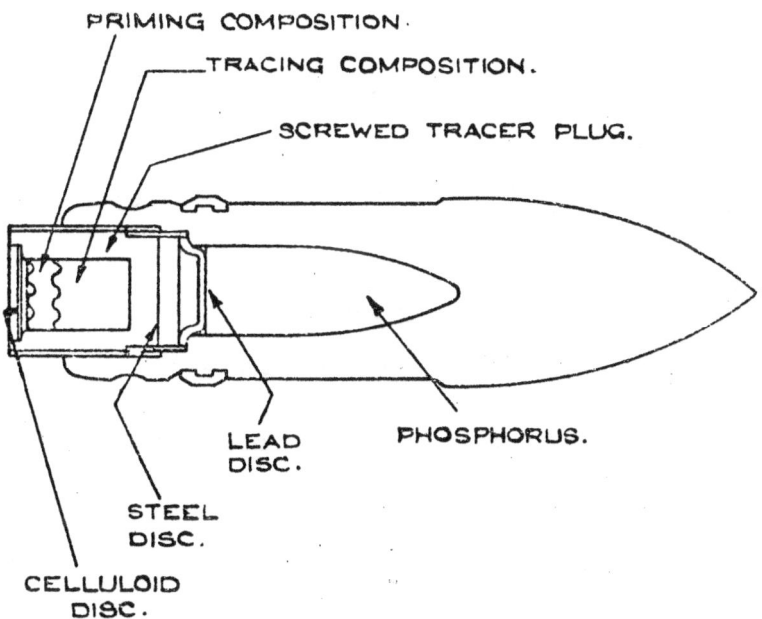

FIG 29
GERMAN 2 CM. SOLOTHURN A.P/T SULPHUR FILLED SHELL

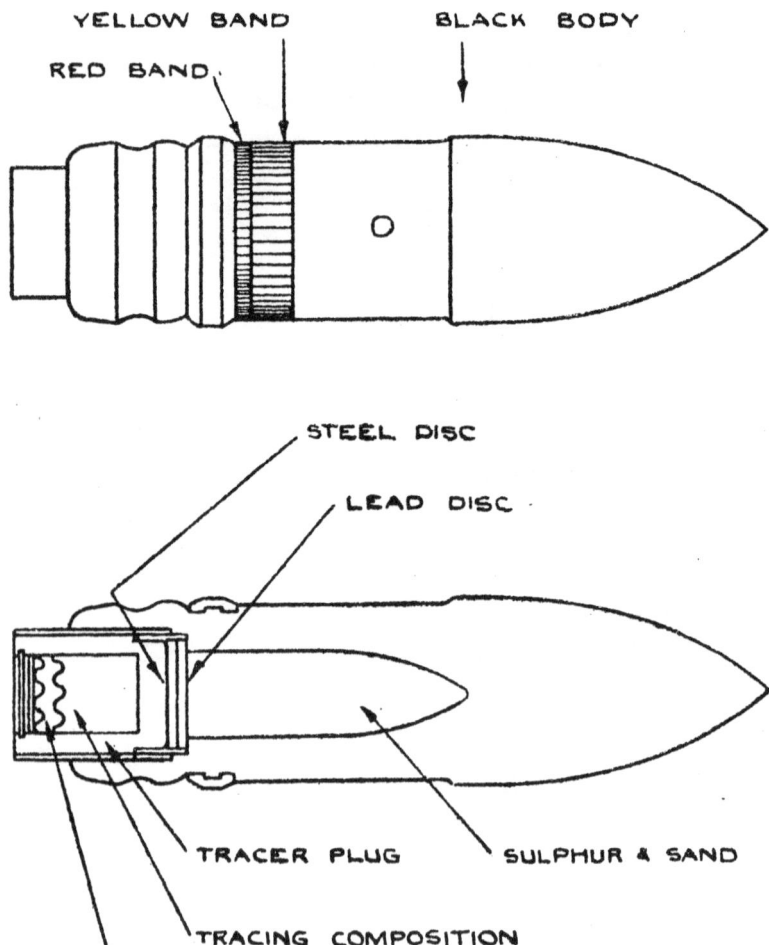

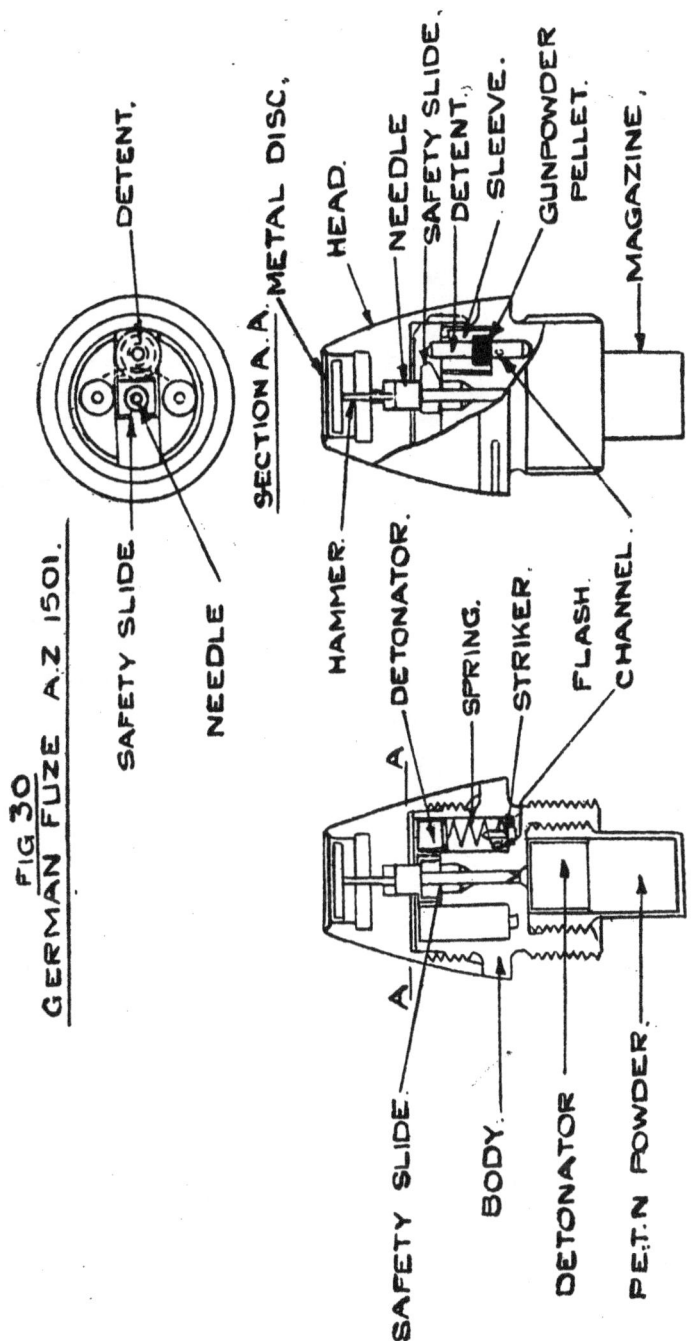

FIG 31
GERMAN FUZE AZ.1504.

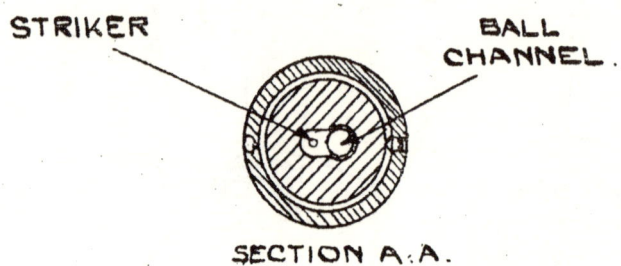

SECTION A.A.

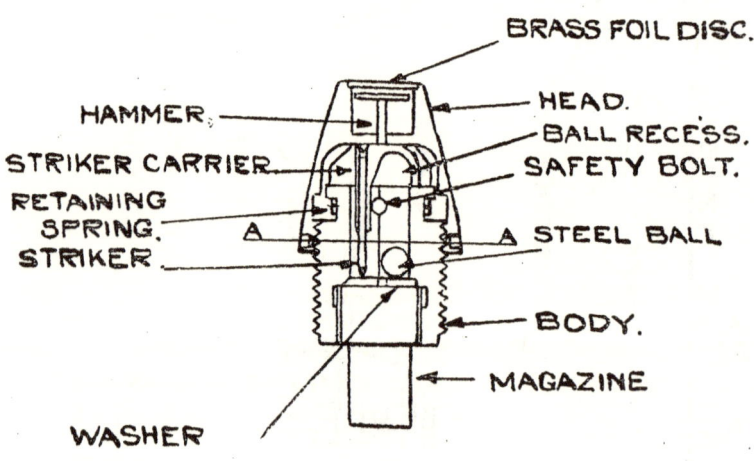

FIG. 32
GERMAN SELF DESTROYING FUZE Z.Z. 1505.

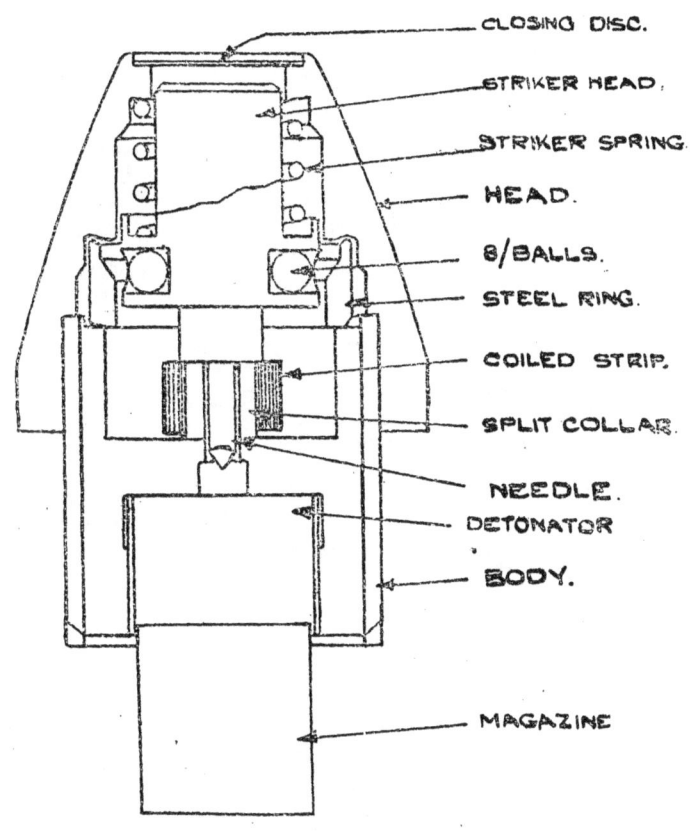

FIG. 33.
GERMAN FUZE Z.Z. 1505 MAGAZINE WITH DELAY FITMENT.

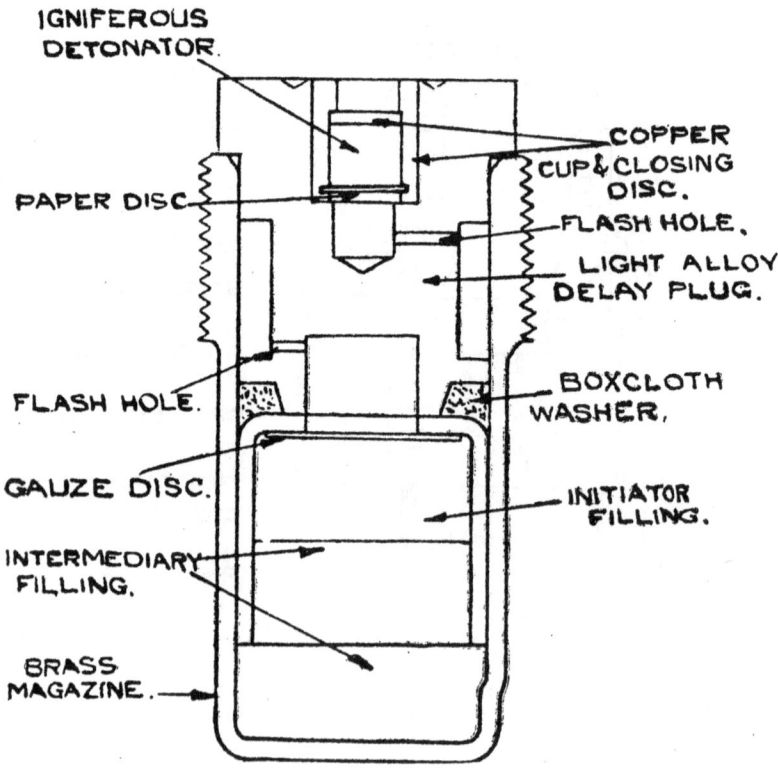

FIG. 34
GERMAN FUZE A.Z.5045.

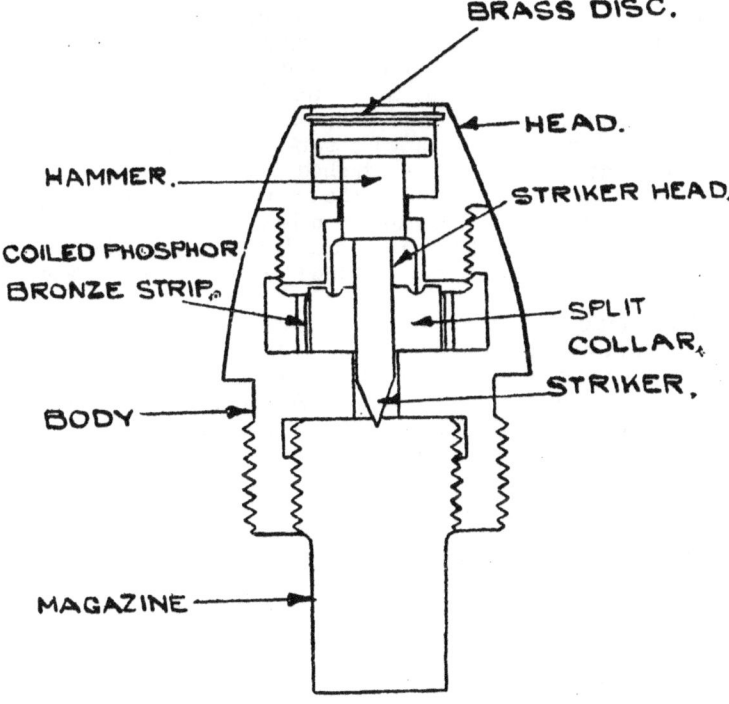

GERMAN EGG-SHAPED H.E. HAND GRENADE
(EIHANDGRANATE 39) (Fig. 35)

The grenade is oval in shape and is dark grey with a bright blue ball cap. The cap is connected by a cord to a friction igniter which is initiated by unscrewing the cap and pulling it with the cord attached immediately before throwing. The weight of the grenade is 7 oz. 8 drams. The bursting charge weighs 3 oz. 14 drams. and consists of ammonium nitrate (80 per cent.), T.N.T. and wood meal. The charge is in the form of a loose powder.

The body is made of thin steel and consists of two portions spun together and closed at the head by a flange formed on the steel bush which receives the igniter and carries the steel pocket for the detonator. The bush is screw-threaded to receive the igniter and is spun to a flange formed on the pocket.

The igniter (Brennzünder 39) consists of a short brass or steel body screw-threaded at the head to receive a square nut and the ball cap and at the base for insertion in the bush. An internal screw-thread is formed in the base end to receive the delay unit. The body contains a friction composition through which passes the pull wire. The lower end of this wire is coiled to provide the resistance to the pull. The upper end of the wire has a loop through which is threaded a two-and-a-quarter inch length of cord. The cord at its outer end is attached to a loose washer within the ball cap. When the cap is screwed to the grenade the slack of the cord is stowed inside the cap. The delay unit consists of a short steel tube filled with compressed powder and is screw-threaded to receive the detonator. The igniter is packed fitted with a moulded transit cap in place of the detonator.

A similar igniter with a red ball cap and a shorter delay (about one second) is used in booby trap devices.

The detonator may be of steel with a screw-thread for assembly with the igniter or an aluminium one of the No. 8 A.S.A. type with a perforated disc pressed in on top of the filling. The detonator is packed separately and is pushed over the delay tube of the igniter when the grenade is prepared for use.

When the ball cap is unscrewed and the cord pulled the friction composition is ignited as the pull wire is drawn through it. The flash ignites the compressed powder in the delay tube which burns through and initiates the detonator.

GERMAN, STICK, H.E. HAND GRENADE
(STIELHANDGRANATE 24) (Fig. 36)

1. The grenade has a cylindrical body of thin-gauge steel containing a bursting charge of T.N.T. and fitted with a wooden handle which contains a friction-type igniter. The igniter is operated by removing the end cap and pulling a porcelain ring attached to a cord at the end of the handle. The weight of the grenade is approximately 17 oz.

2. The steel body contains a 5-oz. 10-dr. bursting charge of loose T.N.T. in a waxed paper container and is closed by a steel closing disc fitted with a detonator pocket. A steel adapter with an internal screw-thread to receive the handle is fitted over the closing disc.

3. The beech handle is hollowed to accommodate the igniter assembly and is fitted with a steel igniter holder at one end. The igniter holder is

screw-threaded internally to receive the brass bush on the igniter. The holder is also the means of attaching the handle to the body and for this purpose is provided with an external screw-thread to engage the adapter fitted to the body. A steel sleeve is fitted over the holder to cover the junction between the adapter and holder. The outer end of the handle is fitted with a zinc sleeve in which a screw-thread is formed to receive a zinc closing cap. A retainer consisting of steel and millboard washers attached to a spring by a rivet is carried under the cap.

4. The igniter (Brennzünder 24) consists of a lead tube containing a copper capsule of friction composition through which the pull wire passes. The upper end of the wire is looped and protrudes. The lower end is coiled to provide the resistance to the pull. A steel tube containing a powder pellet is screwed to the lower end of the lead tube. This pellet provides a delay of $4\frac{1}{2}$ seconds. A brass bush is fitted at the end of the steel tube with a left-handed screw-thread for assembly in the igniter holder. The detonator is of the No. 8 A.S.A. type and is a push fit in this bush. The ends of a length of twine, threaded through a porcelain ring, are led through the protruding wire loop and retained by a wooden bead. The twine passes through the handle to an enlarged space at the outer end where the twine and porcelain ring are stowed under the retainer and screwed cap.

5. The handle, with igniter assembly fitted, is packed separately. The detonator is also packed separately. These components are assembled and inserted when required for use.

6. Before throwing, the screwed cap is removed from the handle and the porcelain ring pulled to draw the pull wire through the friction composition. The flash from the composition ignites the delay pellet which burns through and initiates the detonator.

GERMAN SHAPED DEMOLITION CHARGES.
(FIG. 37)

1. These charges are designed on the " hollow charge " principle and are intended for the perforation of armour plate.

2. Two types are known to exist; the 12·5-kg. (27·5-lb.) charge and the 50-kg. (110-lb.) charge.

3. The 12·5-kg. charge consists of a light metal bell-shaped container which contains the T.N.T. blasting charge with a hemispherical cavity and an intermediary prepared to receive a detonator. The charge container is fitted with a handle for transport.

4. The 50-kg. charge is made up in two parts for convenience in handling, each part being provided with a handle, and one part is superimposed on the other when positioned for use. The containers are both of light metal and are shaped to fit closely together when assembled. The T.N.T. filling of the inner part has a hemispherical cavity while the filling of the outer part, also T.N.T., includes an intermediary prepared to receive a detonator.

5. The charges are positioned with the flat surface on the plate and are initiated by means of a detonator of the normal type which is inserted in the intermediary and ignited by means of a length of safety fuze and a friction igniter.

6. The perforation performance of these charges is reported to be as follows :—

(a) The 12·5-kg. charge will perforate 4·7-inch armour plate.
(b) The 50-kg. charge will perforate 9·8-inch armour plate.
(c) A 50-kg. charge followed by a 12·5-kg. charge, laid on the same spot, will perforate up to 11·8-inch armour plate.
(d) Two 50-kg. charges fired in succession over the same spot will perforate up to 19·7-inch armour plate.

The charge has to be positioned to provide a good contact between its hollowed side and the plate. The hole produced is usually about 3·9 inches in diameter on the exterior surface of the plate and 23 inches on the interior surface.

FIG. 35.
GERMAN EGG SHAPED H.E. HAND GRENADE.
(EIHANDGRANATE.)

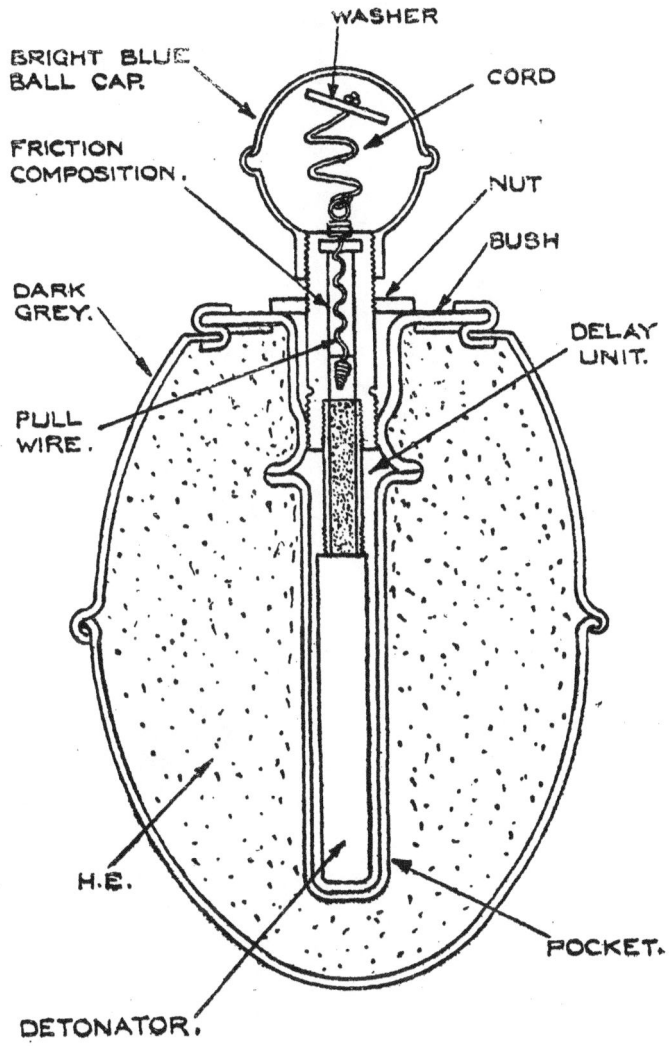

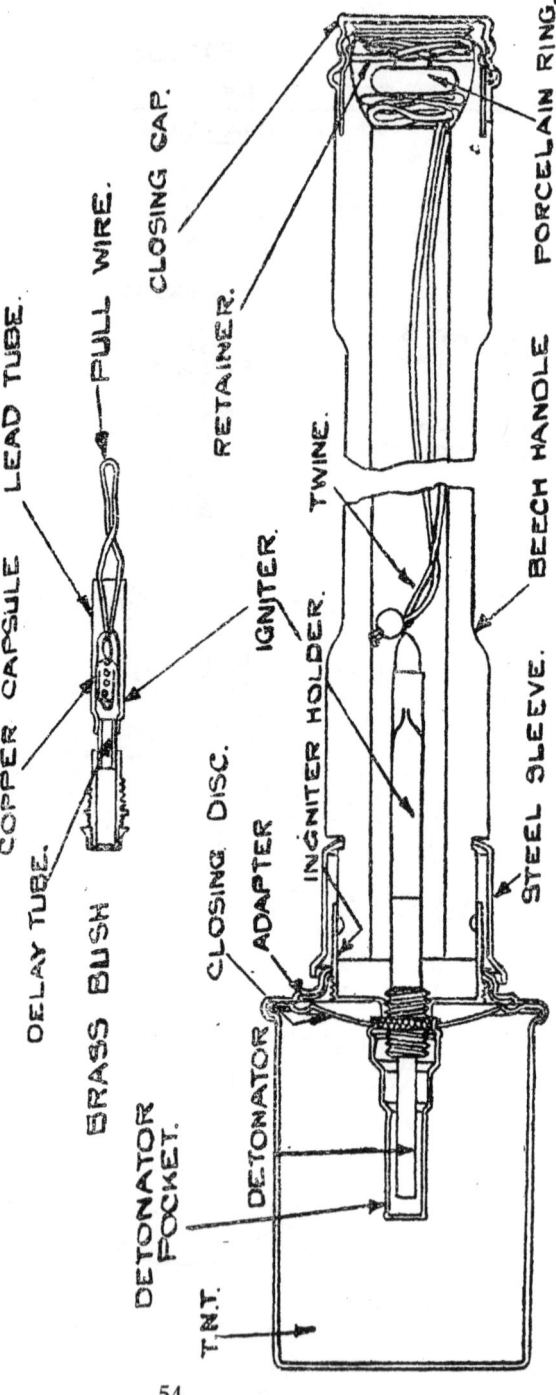

FIG. 36.
GERMAN, STICK, H.E., HAND GRENADE.
(STIEL HANDGRANATE. 24.)

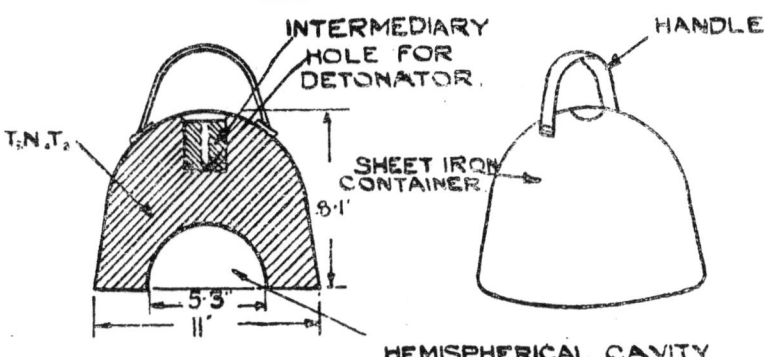

FIG. 37.
GERMAN DEMOLITION CHARGES FOR ARMOUR PLATE.

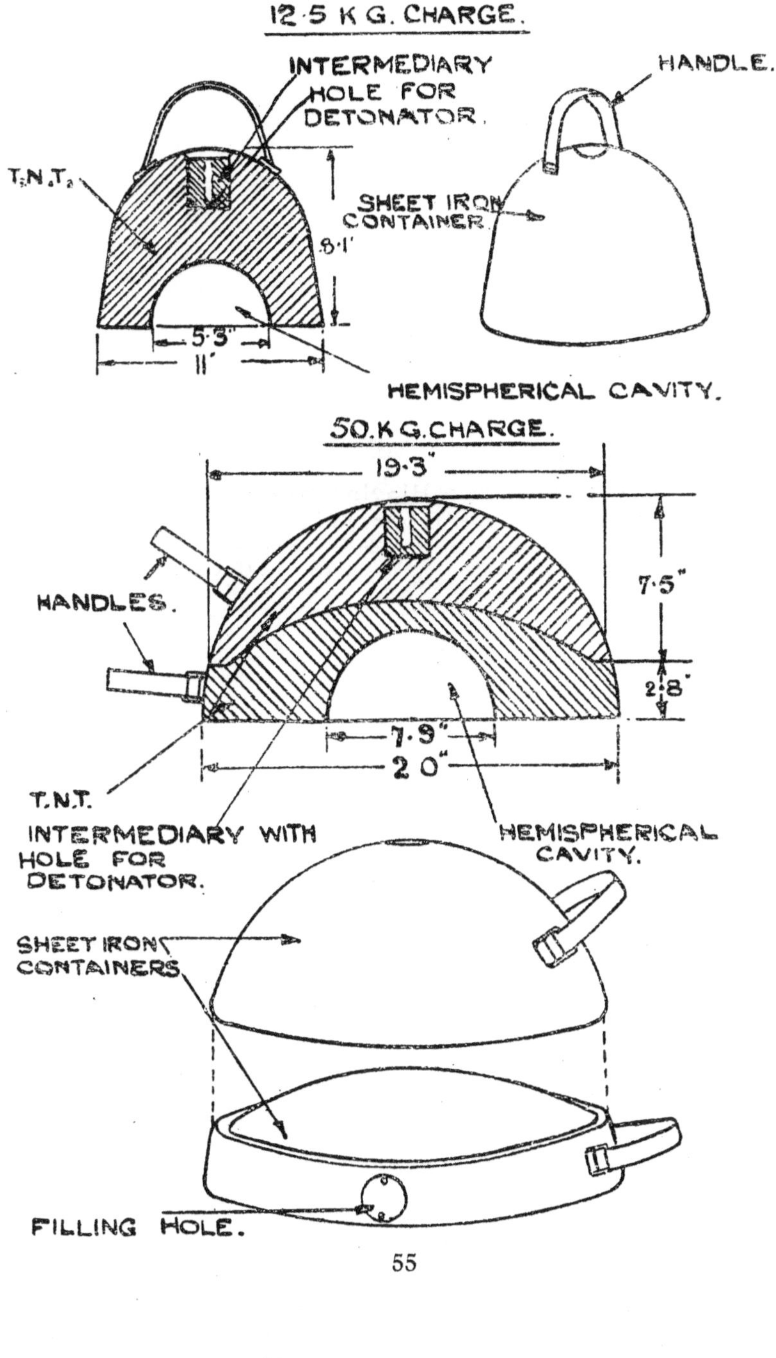

Printed under the Authority of HIS MAJESTY'S STATIONERY OFFICE
by William Clowes & Sons, Ltd., London and Beccles.

(1975) Wt. 33871—9623. 6,000. 12/42. W. C. & S., Ltd. **Gp. 395.**

www.ingramcontent.com/pod-product-compliance
Lightning Source LLC
Chambersburg PA
CBHW032012080426
42735CB00007B/582